the cob builders handbook

You Can Hand-Sculpt Your Own Home

By Becky Bee

This book is dedicated to Mother Earth.
Thank you so much for everything!

Library of Congress Cataloging in Publication Data

Bee, Becky
The Cob Builders Handbook: You Can Hand-Sculpt Your Own Home

Library of Congress Catalog Card Number: 97-90728

ISBN (paperback) 0-9659082-0-8

Illustrations by Becky Bee
Book Design by Becky Bee, Alex McMillan, Mitch Spiralstone
Book Layout by Alex McMillan

Ordering Information
Distibuted by Chelsea Green 800-639-4099
For copies of this book send your address and:
(**within the US or Canada**) US$23.95 this includes shipping costs.
(**outside the US or Canada**) US$29.95 this includes shipping costs.
to: **GROUNDWORKS** P.O. Box 121 Azalea, OR 97410, U.S.A.
Contact us for bulk ordering details and overseas shipping prices.

ACKNOWLEDGMENTS

Special thanks also to: Mitch, Alex, The Yellow Pages Book Club, Sun Ray, Bella and Richard, Ianto and Michael, Billie Miracle, Sequoia, Jill, Jean, Evelyn, The Steens, The Adobeland Galz, The Southern Oregon Women's Writers Group, and all the wonderful people who have encouraged and facilitated the birth and evolution of this book.

Please send in your comments and ideas. They may be included in the next edition.

BE FOREWARNED!

Cob gets under your fingernails, into your bones and deep in your heart! If you build with cob, you will be transformed and you will never be the same!

Caution: Cob is addicting! Becky and all the people who have contributed to this book do not assume responsibility for the financial, mental, and physical health and happiness resulting from the use of this book. It's all yours!

Neither do we assume responsibility for damages, losses or injuries that may arise from the use of the information in this book. Every project and situation is unique. Please use good judgment and common sense. Take care of yourself and Mother Earth.

HAPPY BUILDING!!

Contents

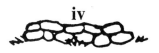

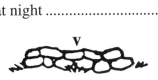

vi

INTRODUCTION

The purpose of this handbook is to show you how you can build your own magical, practical, long-lasting home for very little money and have a wonderful time doing it!

Cob is in the early stages of being rediscovered in the modern world. Ideas and innovations are popping up all the time. I wish I could say I learned cob from the folks of the past generation, but I can't. I am sharing my "modern" cob experience and current thoughts.

Because cob is such a tactile kind of thing, I usually teach about it via hands-on workshops. It's easy to show people how to do it. When I sat down to write this book, I was amazed at how many words it takes to describe something that fingers can

understand without a single word! Cobbing is easy. All the text makes a simple thing seem a lot more complicated than it really is. Try it! You can do it! Think of the words in this book as a reminder of what your common sense and ancient memory already know.

My intention in writing this handbook is to encourage the rebirth of natural building. This book is designed to make it easy for you to join other pioneers in this wholesome adventure. It is written for people with or without building experience. I hope this is the kind of book that you will want to keep and pass on to young people to inspire them to build natural homes in the future. (I wish that it could be made out of cob so it would last for hundreds of years.)

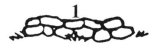

Building with cob is a powerful political action, greatly reducing the need for the mortgage systems, lumber and construction industries, and petrochemical companies. Cob builders spend less of their lives working to pay for all of the above, and more time living. Making homes with natural materials gathered gently from the earth improves the likelihood of the survival of life itself.

Throughout history, women have worked together homemaking, farming, cooking and raising children. This is the glue of community. Today in the modern western world, most women are isolated from one another and are usually dependent on men and/or the patriarchal system for their shelter. Cobbing is a way for women to re-experience a sense of community and be empowered to make more life choices for themselves.

This book is put together by me, Becky Bee. I've loved building as long as I can remember. When I was a kid, I built tree houses and designed underground forts. I grew up in Central America and spent a year in Africa. The beauty and serenity of the natural houses there felt like home to me. When I was living in New Zealand during the eighties, I was excited to find cob homes in the western world.

As long as I can remember I've had the vision of sustainable living - the garden, the handmade house, the little creek. I love the idea of being part of nature.

I've enjoyed creating lots of different kinds of buildings: log cabins, sweat lodges, tents, teepees, straw bale, adobe, conventional frame, recycled wood and cob homes. I was first introduced to cob in 1989 at a workshop offered by Bella and Richard Walker in New Zealand.
In 1993 I took a course in pottery, fell in love with clay sculpting, and found the artist in myself. Around that time, I again went to a cob workshop, this time being taught by Ianto Evans and Michael Smith. I picked their brains and cobbed with them that year. We learned a lot and had a lot of fun. Cob building brought together my loves: clay, beauty, home and building!

I have been researching cob and teaching about it ever since. I love the feeling of being part of a team working together to create a strong, sensuous building. I am absolutely delighted to have found something that I love to do that makes sense, in a world where lots of things don't!

What is cob?

If you would like to skip the introduction and go straight to the chapter on cobbing, turn to page 71.

The dictionary lists one of the root meanings of cob as a 'lump' or 'mass'. One definition of cobble is 'to make'. And a cobber is 'a friend'. So let's cobble a cob house with our cobbers!

Cobbing is a process best described as mud daubing. Earth, sand and straw are mixed together and massaged onto the foundation, creating thick load-bearing walls. It's like hand-sculpting a giant pot to live in.

Earthen homes are common in Africa, the Middle East, India, Afghanistan, Asia, Europe, South and Central America. Easily one-third of the world's population is currently living in homes made of unbaked earth.

The three most common forms of earth buildings are adobe, rammed earth and cob. In the southwestern United States, the five hundred year old Taos Pueblo, as well as many homes and churches, are made of adobe. Adobe is a form of building using unfired earth. Dirt, straw and water - the same ingredients as in cob - are made into bricks which are then sun dried and built into walls with a "cob-like" mortar. Some very old Native American structures like the Casa Grande ruin in Arizona are made out of cob. These are described locally as being built of "puddled or coursed adobe".

There is evidence that cob building began in Europe about 800 years ago. Some buildings that were built in the 16th and 17th centuries are still standing today. In England, there are approximately 50,000 cob buildings still in use today. Most of these were constructed in the 18th and 19th centuries.

Unfortunately, with the advent of fired brick construction, and political alliances between brick makers and the masons, the skill and art of making homes out of cob almost died out in Europe over the last century. Since 1980, the traditional craft of cob building has been enjoying a revival, mostly in the form of repairs or additions to existing buildings, with some new structures being built as well. In 1996 in Britain, four new cob buildings were under construction with building council approval.

Why build with cob?

It's fun and enlightening! It inspires getting acquainted with and connected to nature, oneself and one's fellow cobbers. Building with cob definitely enhances well-being.

Cob is gentle on the planet. Using cob reduces the use of wood, steel, and toxic building supplies.

Getting to know Mother Earth

It's fun to use your mind to figure out how moisture, gravity, heat, seasons, temperature, and water function in nature. Observe how nature creates form and beauty. Notice patterns in nature's structures, in plants, bones, a snail shell, bubbles, a cobweb, a nest, etc.

Get to know your environment so you can treat it respectfully. Notice how your actions affect nature. Gather what you need for building in places where you will cause the least damage/impact. Learn which trees to cut, and at what time of the year. Where does your home fit into nature gracefully?

Cobbing is a fun way to get to know Mother Nature. Building with cob satisfies some ancient human urge and reminds you that you really are part of the natural world.

It's fun being in charge of the creation of your own home.

Cobbing requires defining what it is that you want. This process will give you a clearer picture of who you are! You will be the creator of your environment in every step of the process: designing, building, and decorating!

Cobbing connects you to the long forgotten memories of building with nature that have been stored in your cells and passed down from your ancestors. It will help you **remember that you are a child of Mother Earth.** You'll get to know the inventor in yourself, the artist, the inspired creator, the designer, the organizer, the homemaker. Become familiar with your own ingenuity and intelligence. Stretch your ability to visualize and to answer questions. Cobbing will help you develop confidence in the many facets of your being.

Cobbing is good for your body! Watch as it gets stronger, harder and healthier! You'll learn to move efficiently and pace yourself. Cobbing is rhythmic, slow and constant so your fitness kind of sneaks up on you. I've heard claims of cob healing just about everything from anorexia to arthritis!

Cob building sites are usually quiet and safe places for people to be. Cob structures make nontoxic, healthy indoor environments for the occupants.

Cob helps you get to know yourself in relation to others.

It's easy to inspire others to participate in cob building projects because it's so fun and satisfying. People are happy when they are part of a team making something beautiful and useful! Cobbing lends itself to sharing, co-creation and group decision making. It reminds us that we can still function as a clan.

Cobbing together inspires deep sharing and friendship. Wonderful conversations seem to arise out of the mud! Cob building sites welcome people of all ages and abilities to join in the fun!

IT'S EASY!

You can do it! Cob is a flexible and forgiving medium. It requires dedication more than physical strength, and willingness to experiment more than skills. Building with cob is an easy way to go on a big adventure!

IT'S AFFORDABLE!

Did you think you might have to pay rent forever? Now that you have this book, you'll see there's a way out! If you have a place to build on, and you're a good scrounger, and if you're good at getting people together for house raising parties, your cob home can cost as little as $10 a square foot! A well-designed cob home will save you money on energy bills too!

A conventional house made of concrete, 2x6s, chip board, plastic, fiberglass, vinyl, aluminum, composition shingle will cost at least $65 a square foot. (These are 1996 prices.)

IT'S COMFORTABLE!

The organic shapes of cob walls are pleasing to the eye. **Walking into a rounded cob home is like walking into a hug.**

Naturally cool in the summer and warm in the winter, a cob home will keep you comfy year round. The thermal mass of the walls slows down changes in the temperature. Heat from a fire and from the sun will be absorbed and radiated back to you during the cooler nights. Cob walls also muffle sound, making a quiet indoor space.

IT'S LONG LASTING!

No, cob homes don't melt in a downpour. Cob homes last for centuries. With the revival of cob and other natural building techniques, the next few generations can choose not to spend their lives paying off a mortgage. Cob homes don't require much maintenance either.

YEAH FOR COB!! It's too good <u>and</u> it is true!!

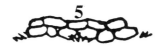

DESIGNING YOUR HOME SWEET HOME!

You can make a house that looks like any other, or you can take advantage of cob's sculptability and make a unique building, one of a kind.

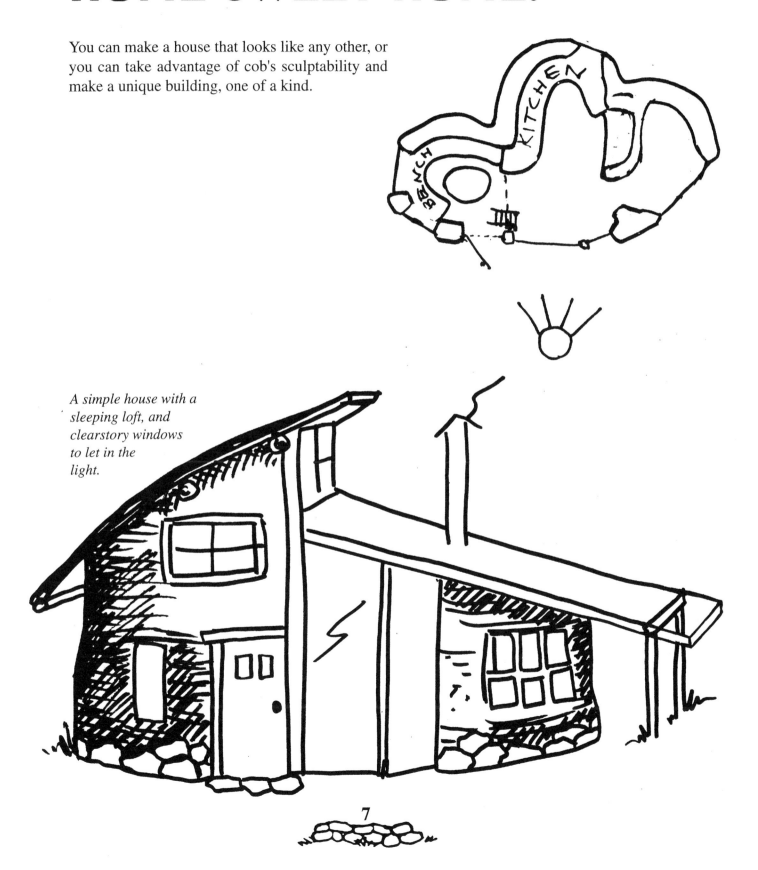

A simple house with a sleeping loft, and clearstory windows to let in the light.

Things to do to get ready

- **Go over this book once or twice** to get familiar with cob concepts. Read the checklist (page 103) and visualize your home.

- **Read books.** At the library you probably won't find much on cob. Read up on other types of earthen construction and natural building. Books on permaculture, foundations and drainage, stone wall building, house design, passive solar design, electricity and plumbing, and roof construction will be helpful. (See the recommended reading list at the end of this book.)

- **Go to a cob workshop** if at all possible. Getting your hands in it will give you lots of information, inspiring ideas, confidence and rekindle your cellular knowledge. Workshops are a good way to get familiar with the material you will be building with.

- **Observe, define and document what it is about different environments that appeals to you.** Visit buildings that feel good to you. Check out the natural buildings in your area. **Collect your ideas, sketches and photographs in a scrap book.** Include the ideas you come up with while reading this book.

- Take design **lessons from Mother Nature.** Spend time in nature noticing how she puts things together. What shapes appeal to you? Colors? Textures?

- **Assess your resources:** What do you have in the way of time, energy, money, materials, skills and helpers? If you decide to have workshop(s) and/or house raising parties, set the dates and start advertising. Get your friends inspired.

- **Use your imagination as you design.** Keep in mind the feelings you would like to evoke in your home. **Be flexible** about your design. Once you've established the foundation, the spirit of cob will help you design from there up.

- Design your home so it **belongs** where it is. **Spend lots of time on your site during the design process. Plan any roads, parking, energy systems, water sources, the garden-orchard areas, etc., at the same time, so that everything will work together gracefully. Include the sheds and storage areas in the design.** It is important to **make sure the home site will be as dry** as possible. Plan a drainage system that moves water away from your house. (Read the chapter on choosing a site for your home carefully! See page 17.)

- Start **gathering materials**: roofing materials, doors, windows, etc. These elements will influence the personality of your design.

- After you've started forming some clear ideas, **make small cob models,** ideally right on the future home site. The models can be made of potter's clay or cob. This exercise is incredibly valuable and will teach you a lot about your design ideas. You can mock up your design full-scale with stacked straw bales too.

- **Allow plenty of time for the designing and building processes.** A home full of love is made by people who take the time to love the process.

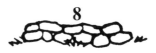

Think Small

Building a house (even a small house) is a **big project!** It's best to start something that you can finish without too much stress. You can always add a room on later. If you become a cob addict you can add on each year.

Smaller buildings usually require wood of smaller dimensions to support the roof. This reduces the need to cut trees and spend money.

A well-designed small building is all a person needs. It's cozy and easier to maintain. Measure the rooms where you live now so you can quantify the sizes of spaces.

A small home encourages more time spent outside in Mother Nature. **Design outdoor living spaces around your home.** Patios, covered porches and doorways that invite you outside add a lot to a home.

Design your home around what you intend to do in it. Write down your daily activities and think about what time of day you're likely to do each thing. Take advantage of natural light and heat when the sun shines into different parts of the house. Make a rough sketch using circular shapes to represent your daily activities. This will give you a basic layout for your design. Try arranging the layout in a variety of ways. How can one place be used for more than one purpose? Plan the flow of people-traffic carefully.

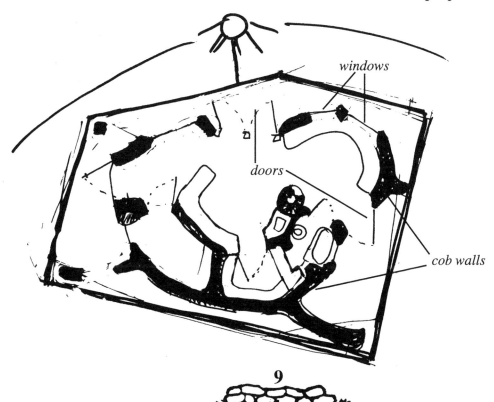

windows

doors

cob walls

Remember to include **lots of storage space** (at least 15% of the floor area). A small house feels bigger when it's tidy and not cluttered. Design in lots of closets, shelves, niches, and hooks. Plan a shed off the house for wood and tool storage.

People almost never stand right next to a wall. This means the walls don't need to be as tall as you are, if the ceiling is slanted/ sloped. Be willing to let go of what you're used to while you are imagining your home. The lower the walls, the less work you'll have to do to make them, and the less energy it will take to heat your home.

Think about **efficient use of kitchen space**. An old wive's tale says to put your sink, refrigerator and stove in a triangular relationship to each other. Old wives probably know what they're talking about. Notice kitchens. Which ones are comfortable and efficient? What makes them that way? Copy them. Arrange a visit to see the inside of a yacht or houseboat to get some good ideas for kitchens designed to take up a minimum amount of space.

It's easy to make beautiful furniture out of cob. **Furniture that's built against the walls takes up a lot less room** than movable furniture and leaves the floor space more open.

Think Rounded

Now is your chance to let go of the straight, square concept of home. Observe nature. She rarely uses a straight line and her graceful structures have survived many tests of time.

Curved walls are more stable than straight ones. The tighter the curve, the stronger the wall. A long straight wall wants to fall over. A curved wall holds itself up.

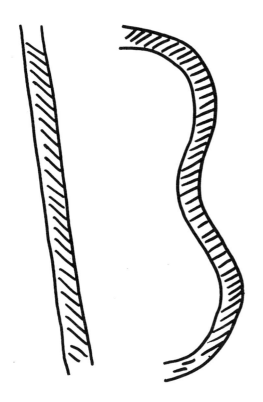

If you must make a long straight wall, add a buttress or two. (See page 13 for more on buttresses.) When a cob wall curves tightly, you can deduct a few inches off its thickness because a curve is so strong. Where a cob wall is long and straight, increase the thickness of the wall.

As you imagine the walls, also start visualizing the roof and how it will sit on the building. The roof design can be refined as you build.

A wall defines the space on either side of it. Generally people are more comfortable in spaces with positive angles. (More than 80 degrees and less than 180 degrees.)

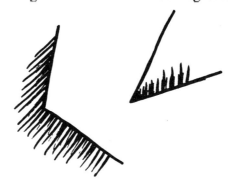

This is easily done for more than one room if the rooms are all squares or right angles, but it's a little trickier if you want rounded walls. An easy way to see how nature solves this dilemma is by looking at the honey comb of a bee hive, or a cluster of bubbles that are sitting on a flat surface. Each wax cell or bubble represents a "room" that is made up of comfortable angles. This also demonstrates a very efficient use of space, maximizing the size of each room in relation to the amount of surface area (or walls).

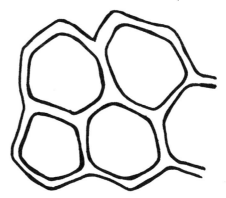

A partial wall, like a counter or built in furniture, is enough to create the feeling of a comfortably shaped room or space.

Keep this concept in mind while you are designing the outdoor spaces around your home too. Take into consideration other buildings, fences, trees and the outdoor terrain. These too, will define the space.

Make the Most of the Climate
Design with passive solar access in mind

If you live in a temperate climate where the sun shines during the cold months, use the heat of the sun to heat your home. Using passive solar design means you'll use less energy, money and fuel to stay warm. Let the sun shine in! This is an important part of designing your home. (Read the sections on passive solar design on pages 58 and 114-117 very carefully.)

cob is thermal mass

This means it is dense stuff that holds the heat from the sun or the fire for a long time, radiating it back out into a room slowly. It also takes a long time to heat up. In most climates where humans live, this helps create a pleasant indoor temperature. Cob is naturally cool in hot summers, and can absorb solar warmth in the winter. It tempers the climate throughout the year and does the same on a smaller scale throughout the day/night cycle. Cob is ideal in desert areas where it holds the night's cool throughout the day, slowly heating up to release the day's warmth at night. Pay attention to the weather in your area through the seasons. The more sun you get during the colder months, the more practical a solar designed home will be.

If you live where it's really hot or cold around the clock, you may want to insulate or 'outsulate' your cob walls with something that is full of air pockets, and synthetically regulate the temperature by heating or cooling the interior. Adding more straw, pumice, vermiculite, or even styrofoam packaging chips to the outer or inner layer of cob may increase the insulation value.

Other things to think about when designing

One Story or More?

*   A two-story house is more economical and efficient because the foundation and roof are the most expensive energy consuming parts of building. When you increase your living space by building vertically, you use fewer materials in the roof and foundation.
*   Heat goes to the upper story. If you are like me and you like to sleep where it's cool and where it's easy to get to the bathroom, you might want to put the sleeping place downstairs.
*   If you have plumbing it will be easier and quieter to keep it all on the lower story.
*   You'll need lots of lumber for constructing the floor of the second story.
*   Stairs take up more space than you think on the lower level.
*   Stairs or ladders can be hazardous and difficult for the old and young.
*   It is trickier to get a multi-story house to look like it belongs in its environment.
*   It's more awkward and dangerous to work further off the ground.
*   The walls will have to be thicker at the bottom to support a second story. This means you'll need more cob and more foundation.

Noise

Cob walls do an excellent job cutting noise, windows less so. Hopefully the noisy side is not the sunny side. Design accordingly.

Plan for Future Additions Now

If you plan to add onto your home in the future, it's important to design the different stages so they complement each other well. Keep future addition(s) in mind as you design. Make sure the water will run straight off every roof into a gutter, and not onto another roof. (See page 38 for more about future additions.) You may want to consider a C or L or S or U shape for modular building.

Designing the Entrance

The entrance to a home is a big part of its personality. It creates an impression for everyone as they come and go.
It's worth investing thought and imagination

into this important aspect of designing your home. (See the section on designing the door area, page 36.)

Buttresses

To strengthen a wall design, either because you want a thin wall or because you've decided to make a long straight wall, plan in some buttresses to help support it. A buttress is a secondary construction that adds lateral support to the wall.

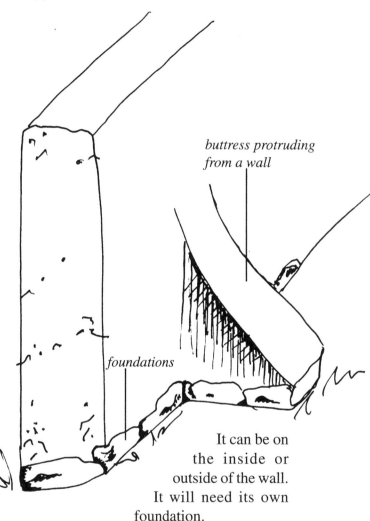

buttress protruding from a wall

foundations

It can be on the inside or outside of the wall. It will need its own foundation.

Interior walls that join the exterior walls serve as buttresses. Any substantial blob of cob, like a fireplace or furniture against the wall, will give the main outer walls lateral support. Furniture and interior walls will support less weight than the main walls so they can have less substantial foundations.

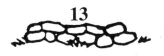

Interior walls take up precious space so the thinner you can make them, the more space you'll have.

Ideally, you'll plan the buttresses as you plan the home so they can be built at the same time as the main walls. This will make a strong connection between the main wall and the buttress and its foundations.

If you build a buttress on the outer side of the wall, it must be protected from the rain - either with extra long eaves or with its own little roof. The same is true for outdoor garden or patio walls.

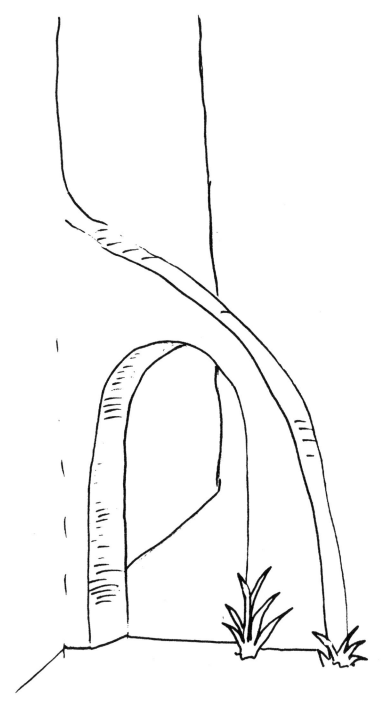

Flat stones or tiles can be carefully laid right on the top of the buttress to make a roof.

A big buttress with a door in it makes a romantic entry to the garden or courtyard.

Permits

If you live where building permits are required and you want to get the **building permitted, you will probably run into some resistance.** Some commissions don't require permits for buildings under a certain size. Many counties will permit a pole structure and won't be too fussy about what you use to fill between the poles. The New Zealand planning commission has recently drawn up the requirements for building cob houses. They still require an engineer's approval before the building permit is granted. There's lots of work to be done before cob becomes an acceptable building material in most parts of the western world. Good luck! When dealing with building officials it is helpful to think of them as your friends and ask for their help.

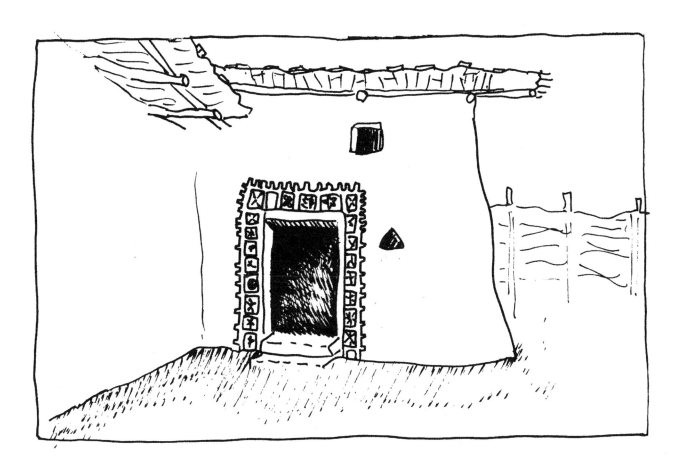

CHOOSING YOUR SITE

You know what they say: the three most important things about real estate are location, location, location! It's true for your cob home too!

Here's a list of some things to consider when looking for land:

- Getting land to build on is the first and probably the biggest step towards your dream of "home sweet home". Reading this book will give you some good ideas about what to look for. Take your time, use your intuition and be brave! If you choose to lease or rent land, make sure you have a very clear, written legal agreement with the landowner that ensures your right to live on the land.

- **Who** will you live with? Are you sure you want to live with this person(s)? Make clear agreements regarding the use of the land with your land partners.

- Your **feelings:** do you love the place? Are you ready and willing to make a **commitment** to this piece of land?

- **Restrictions**: building codes, covenants, zoning, minerals, water, and access rights

- **Future plans for the surrounding areas:** clear cut? a non-organic farm? noisy factory?

- **History** of the land

- **Economics**: price of the land, payment plan

- **Climate**: solar access, rainfall, wind, potential disasters: earth-quake, hurricane, flood, etc.

- Good **homesite(s),** drainage

- Dependable year round **water** source for drinking and irrigation

- **Erosion**: flood plains? clear cuts uphill of the land?

- **Neighbors**: privacy, noise, property lines

- Surrounding **community**: diversity, culture, schools

- **Views**

- Toxic **pollutants** in the area, or on the land

- Can you get what you need and want in a nearby town? resources, **services**, jobs, economic health of the area

- **Accessibility:** check out economic and environmental costs of road building and maintenance, commuting **distance** to town, road conditions

Finding your home site

Once you get land, you can start deciding where your home will be situated on it. It's a good idea to allow lots of time for this important step. Observing the land in all the seasons can be extremely helpful! **Spend as much time as possible on the land**. To make it easy for you to hang out there, set up a little camp kitchen, shelter, hammock, etc.

Read the design section of this handbook while you're considering your site options. Also read the section on drainage to help you choose a spot that will minimize your drainage work. It may be helpful to read up on permaculture, which is a design system that considers the multi-facets of life in the planning process.

Choosing the house site includes:

- Choosing the **water source** and planning how to get it to the house.
  Will you be using water from the county system, a well, a spring or a stream? Will it get to your house by gravity feed or will it be pumped? Will it flow all year?

- Designing the **septic or waste system.** Generally the simpler the system, the better. There are lots of books about these subjects, everything from outhouses to by-the-code plumbing.

- Visualizing the **access and parking.** Building roads is one of the most destructive things people do to Mother Earth. Roads are often the cause of erosion and landslides, so plan them very carefully. Designing a road is complicated and can be expensive. Take the time to learn as much as you can about it. Do not assume that the guys you hire to do the road know what they are doing. Keep an eye on the road during heavy rains. Take your shovel out and adjust the road where necessary to protect it from erosion.

  Do you want to see approaching cars? Do you want them to see you? Cars are noisy, stinky and usually pretty ugly. I suggest keeping the parking lot out of your view as much as possible. The approach and entrance to your home will influence your home's character.

  It's very convenient to be able to drive a load of sand or rocks right up to the homesite during construction, so you may want to at least make a temporary road for that purpose.

- Laying out the walking **path(s)** to and from the house, garden, outhouse, etc. These need to be practical and direct. Design them with surface water runoff in mind. Use the same strategies as for making a road. Walking paths can easily become creek beds!

- **Analyzing the soil.** Where is the best soil for plants? Where is it naturally the most stable and dry for building? What is a practical route for a road? Where can you find the best cob mix? (Read the Cob Glorious Cob Chapter for more about cob mixes.)

- Finding a place for your **garden and orchard.** Do you like your garden to be close to the house? If you can see it from indoors through a window, it invites you to spend time in it. If there's critters to keep out, you'll need to build a fence. If you think fences are ugly, you may decide to put the garden where you can't see its fence from the house. Or make a pretty fence that you'll enjoy seeing from indoors.

  Placing the homesite near a fertile garden spot on your land will save you a lot of soil improvement work.

  Unless you live in Eden, you'll probably need to water your garden. Think about the water system when you choose the garden spot. Will you be using the runoff from your roof for the garden and/or orchard?

  You may want a road to the garden. If you can drive truckloads of composting materials right to the spot, you will save yourself many wheelbarrow trips.

Put the house where it belongs

A home that suits its environment is a joy to the heart. Choosing a site and designing your home are intricately interwoven. Becoming familiar with the land will inspire your design. Forget what conventional houses look like and let your creative imagination run free. **Let the design grow out of the place as much as possible.**

Pick a site that is naturally comfortable

Pretend you're an animal living outdoors. Find the coziest spots on your land. Where does the sun shine? Notice the winds. Cold sinks to the lowest places and flows over the ground much like water. Where will the cold air sit? Where will it flow? Observe the land carefully in all the seasons. Go there in the biggest storms and on the hottest days. Consider any natural disaster potential like fire or flood and avoid high risk places. Remember that if you put your house on your favorite spot, your favorite spot will be gone.

A dry place is good for the health of your home.

It is important to keep any home as dry as possible. **Choose an already naturally dry spot** such as a rocky outcrop, or a little rise or ridge.
Avoid low areas that will hold the damp. Avoid places where water-loving plants grow, eg. ferns or horsetails. Observe the land carefully during heavy rains. Talk to the former owners and/or neighbors about what happens during high water or flooding. In the wet season, dig some two foot deep test holes on your proposed sites to see how well those areas drain. If the holes fill with water, you'll either want to choose a drier site or create a dry island for your home. (Read the Drainage chapter for how to do this.)

Passive solar planning

Carefully read the sections on passive solar design in the Windows and Doors chapter (pages 114-117).

If you live in a tropical climate, choose a shady, breezy place for your house. If you live in a temperate place where the sun shines during the cold months, **catch the sun to heat your home.**

To find the approximate direction from which the sun will shine, stand on your possible homesite, face the sun at noon, and hold out your arms at right angles to each other.

That's roughly it! The area you're looking at between your hands is where the strongest sunshine will be coming from. Does anything obstruct the useful sun? If there are substantial things in the way like hills or mountains you will probably want to move the home site.

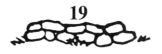

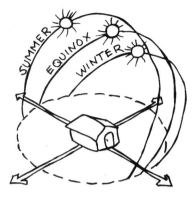

The sun travels high in the sky in the summer and lower in the sky in the winter. The further from the equator you live, the lower the path of the sun will be in the winter. You can find out the exact angle of the sun in different seasons from charts in passive solar books.

You may want to position your home so that deciduous trees can shade it in the summer, or plant some so they'll grow as soon as possible to keep you cool in the hot season. When they lose their leaves in the winter, sunlight can pass through them to light and heat your house. It's wonderful to see the fruit forming and ripening right outside your window.

SUMMER

WINTER

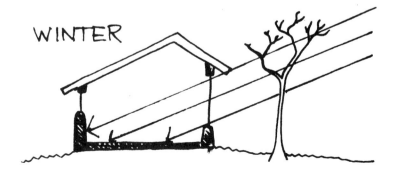

Where the land is protected from grazing animals, the forest will grow back. Think ahead. What used to be a clearing can turn into a dark, damp site. If this is the case where you live, you may need to cut down the baby trees and the undergrowth to keep your site drier and more open, and to keep roots from weakening your foundations.

Think ahead about any evergreens on the sunny side of your home site. They will grow and block your precious sunshine. Either move the home site or consider cutting down the trees. Are these the trees that will provide your lumber/firewood needs?

Harvesting your own wood

Trees can be used for beautiful, round rafters, poles or posts. If you want poles for building, thin the forest intentionally or clear trees from the site. Skin them as soon as they are cut. The fresher the tree, the easier it is to peel the bark off. You can get a special tool for this job called a draw knife, but a hatchet and/or sharpened shovel work fine. Dry the wood in the shade up on blocks to keep it off the ground. The bigger trees can be sawed into boards for roof sheathing, ceiling, and whatever else you want milled timber for. If you don't have a mill, you can get someone with a portable one to come out to your land and do the job. If you want to have the wood cut up at the mill, they will be able to advise you on how to transport and dry the boards.

When choosing a site, consider the trees in the area carefully. Tree roots will grow and can weaken or even destroy a foundation. Any roots below the foundation will have to be removed.

Wind

Do you want the wind blowing on your house? Find a spot that suits you. If it's a cold place, you'll probably want a wind-protected spot behind trees or natural terrain. If you want to plant a windbreak, the sooner you do it, the sooner it will grow. If you live where it gets hot, breezes are an asset for ventilation. (See the section on ventilation, page 118.)

Noise

When choosing your site consider the noise levels in different areas of your land. Night is the best time to really hear noise.

Boundaries

Before you look for a home site it is **very important to find out how close to the neighbors' boundaries you are legally allowed to build.** In some places it can be as far as 200 feet!

GETTING THE SITE READY TO BUILD!

Now that you've put all these factors into the amazing computer on top of your shoulders and have come up with *the* place, it's time to prepare the site for action. Congratulations!!

- Clear trees and brush that are on the site.

- Develop the access.

- Get water to the site.

- Dig an outhouse or figure out a waste system.

- If there isn't a house on the land, set up a temporary shelter and a fire, a cooking set-up, a hammock, and a tent site. Make it inviting to hang out on the land.

- Gather the tools you'll need and make a dry, safe spot to store them.

- Set up a tarp over the construction site for shade and rain protection. This sounds easier than it is. The tarp roof needs to be designed with care. Make sure the water will not run onto the walls or puddle in the middle and pull down the tarp with its weight. If you decide to build the roof of your house first, obviously you won't need the tarp covering.

- Get electricity in if you want it.

- Gather materials and get them to the site. Put everything in the most convenient place so you won't have to move anything more times than necessary. Keep the wood and straw dry. Store glass carefully. (See pages 23-24 for a list of materials to collect.)

- Have a site celebration and blessing.

- Start the drainage and foundation. Yippee!

GATHERING MATERIALS

Remember, for thousands of years people have used what they had and what they could find to build their homes. What follows is a fancy list of all the things you could possibly want while building. The more of this stuff you collect before you start construction, the fewer times you'll have to stop once you get started.

Essential stuff

- a reasonably **healthy body or the power of persuasion** and friends with reasonably healthy bodies

- a fairly **determined, flexible brain**

- **friends** to help

- **water** for cob and a way to transport it: hoses, nozzle, faucet (tap), buckets, water source

- **drinking water**

- a **place to store** things from the elements

- stones for **foundation** or whatever you're using for a foundation (like tires, broken concrete pieces, mortar, forms if you're pouring your foundation, etc.)

- gravel and perforated pipe or tile for the drainage ditches

- **sand, clay and straw** (See how to make cob, starting on page 78, to estimate how much you'll need of each.)

- a **vehicle**, ideally a pickup truck, or at least a friend with one

- little **tarps,** approximately 7x9 feet or bigger, for mixing cob

- **cobbing tools:** squirt bottles, sticks and/or stones to massage the cob together, machete, meat cleaver, and burlap bags or tarps to cover the cob (See pages 51 and 77 for more specific tool lists.)

- big bits of **wood to span** door and window openings if you are not arching the cob over the openings (See the section on lintels page 107.)

- **roofing stuff:** wood, rafters, sheathing and insulating stuff (See roof chapter beginning on page 123.)

- windows, a modern "necessity"

- door(s), and wood for door frame(s).

- step ladder or two, 6 foot or taller

- lots of big **buckets,** multipurpose

Other useful stuff

- **carpentry tools:** hammer, nails, saw, square, pencil, tape measure

- leveling tools: level, taper wedge for the level, long straight 2x4, clear plastic tube for the water level (See page 60.), string

- **earth moving tools:** wheel barrow, shovels, pick, hoe

- plumbing and electrical stuff if you plan to have them, and pipe to run through the walls or foundation

- stuff to set up an **on-site kitchen** if there isn't one to use near by, at least coffee and tea making stuff

- big tarps to cover the site, (or build the roof first)

- string

- rope for putting up tarps

- scrap wood for bracing and scaffold supports, planks and extra straw bales for scaffolds

- sill materials: flat stones, brick, tile

- Piles of lovely things on site while you build will inspire the artist in you. pretty things to bury in walls: hooks, shelves, rocks, tiles, seashells, magic things, colored glass, and colored bottles

- gracefully shaped wood for hangers, hooks, curtain rods, decorations, shelves

- **plastering stuff:** swimming pool trowel, mortar trowel, smooth burnishing stones, pretty colored clays, manure from grass eating animals, pigments, charcoal for black pigment

- hydrated builders' lime if you want to whitewash

DRAINAGE

Drainage is one of the most important aspects in the longevity of your home, so give it plenty of attention. Water can be very bad for your house. (See page 19 for how to choose a dry place for your home.)

The object of the drainage system is to **divert water away from your structure.** You will be **creating a dry island for your home** to sit on. In some areas, a berm will be all you need to redirect water. In others, a French drain and/or a berm will redirect the water. There are more details about these drainage systems late in this chapter.

If you have the time, there are advantages to starting the drainage before building. While you are making your house, you can observe how the drain is working and adjust it if necessary. It will lessen the risk of flooding during construction. And when you are finished building, you will be able to sit down and relax sooner in your new home! If your weather and timing is such that you are unlikely to get flooded out while building, the drainage can be done after you've finished the house.

Test Holes

The first part of the drainage job will be to get to know your land's natural drainage. **In the wet season, dig a few test holes where you imagine your house will be, as well as digging some holes uphill of it.** Do the holes drain well or do they fill up with water? This will give you a good idea about what's going on under the ground and how much you'll have to do to divert water. A sandy, pebbly soil will let water filter through. A soil with high clay content will expand when wet and block water from percolating through it. (Make a little fence around the test holes or cover them with flat rocks or plywood to avoid twisted ankles.)

Making Your Drainage

You may want to use earth-moving equipment, if you have a big job or if there's equipment there anyway for the road building. If you do get machines in to do work, watch them very carefully! Their drivers probably do not have the same familiarity or respect for the land that you do, and can do a lot of damage in a short time. Digging by hand is usually more accurate and less destructive to the land.

The flooded homes I've seen were caused by poorly planned earth-works, human interference upstream (damming, clear cutting, destabilizing earth), or by building where it has always flooded.

Mark where your foundation will be with straw bales, stones, or wooden stakes. Step back and look at the site from a distance. Imagine where the underground water and surface water will flow.

On a sloped site, create a drain and/or berm starting uphill and flowing around the house. (See illustrations next page.) **Your drainage system will direct water to the sides and downhill of your home.**

Consider how you can make use of the water you divert from your homesite (a pond, garden, trees etc). At least make sure it isn't causing erosion. If you won't be catching the water from the roof gutters for drinking or gardens, you might want to channel it into the drainage system too.

Creating drainage on a flat site

Avoid building on level sites if possible. They are better for gardening and much more challenging to keep dry. Obviously there has to be somewhere for the water to go, so **if you must build on the flat, you'll have to create a place for water to flow to.** Make a ditch all the way around the house, draining into a big hole filled with round river stones and/or pumice. This will hold the water and let it soak slowly into the surrounding ground, away from your house or into your water-loving garden planted on top of the drainage area.

In Eastern Nigeria, people solved the problem of building on low, flat, wet land by piling dirt into raised platforms and making their cob homes on top of these, out of the wet. If you do this, remember to tamp the dirt often as you build up the platform.

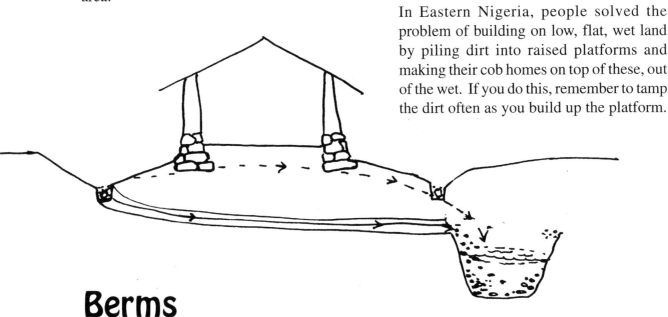

Berms

To divert surface runoff, shovel dirt into a narrow, long, hill-shaped form as shown.

You've just made a berm. Make sure the edges extend far enough beyond your structure, so that the water leaving the berm and flowing freely down the hill will miss your building. You might want to direct the water from your berm to nourish your garden or orchard. If you decide to make a drain and a berm, put the berm uphill of the drain.

For more information about drainage, go to the library and read up on as many methods of drainage as you can find.

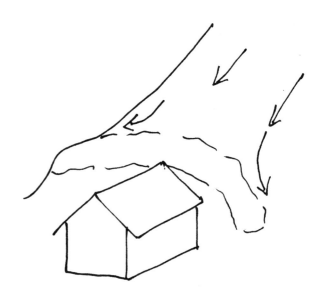

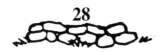

Ditch drains

For wetter climates and underground water, **make a ditch drain.** After making the ditch, you may want to observe the drain functioning under wet conditions before you fill it in. If so, set up walkways so people and wheelbarrows can get across easily, and put up barrier fences so no one falls in.

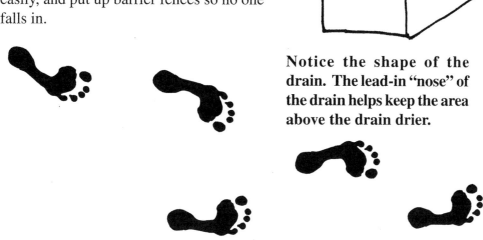

Notice the shape of the drain. The lead-in "nose" of the drain helps keep the area above the drain drier.

How deep do I make the drainage ditch?

Test holes and observing the results after you've dug the ditch will help you see whether you've gone deep enough and whether it flows all the way. (Does the floor stay dry when it's rained a lot?)

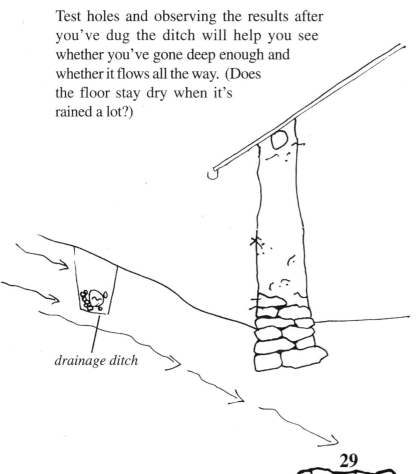

drainage ditch

Your test holes will also give you an indication as to how the water flows under the surface. Ideally, the bottom of the drain will be **lower than the bottom of the foundation.** (See illustration page 55.) The deeper the drain, the safer you are, and the more gravel you will need to fill it. If your drainage ditch is quite a way uphill from your home on a steep slope, it's impractical to make the ditch deeper than the foundation. You'll have to use your judgment about how deep to go to catch any water that might otherwise end up at the house. You might get away with simply making it deeper than the foundation in relation to the surface of the ground.

29

Slope of the ground

For proper drainage, you will need to **slope the ground level away from your home in every direction.** To achieve this, dig on the uphill side of your homesite until you've created a slight downhill angle from the house, then dig your drain at the bottom of that slope. You will be doing three things at once: creating drainage, making a place for a walkway or patio, and getting dirt for building. While you're digging, put the dirt somewhere handy for later use. (Hint: Inside the building is a convenient place to mix cob. Leave enough room to maneuver a wheelbarrow.)

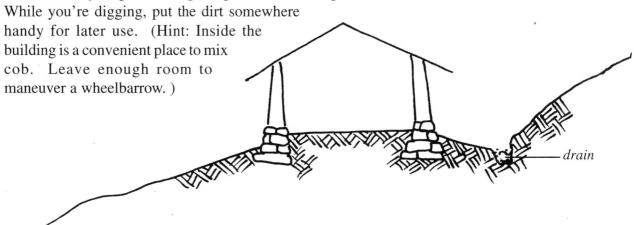

— *drain*

How far is the drain from the foundation?

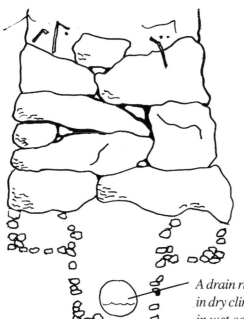

The drain can be right under the foundation, or it can be anywhere from 1-10 feet or more away. If you want to leave the drain open while you're cobbing, it'll be more convenient if it's far enough from your house to be out of the way while you are building. Make sure the water will flow from the house to the drain!

A drain right under the foundation will be enough in dry climates, or can serve as a secondary drain in wet conditions. It also reduces the chance of movement caused by frost.

How wide do I make the drainage ditch?

It needs to be wide enough to fit a 4 inch perforated pipe and 2 or 3 inches of gravel on either side of the pipe. You will need enough room to get your arms in to lay the pipe. (See tips on page 31.)

Completing the drain

You can either do this step before you start the house, or wait until you're sure the drain will function like you want it to, after a series of heavy rains. If your soil has a lot of clay, sliding a shovel along the sides of the ditch will make the clay smooth and slick. This can harden and create a water resistant barrier that blocks the water from entering the drain. Try to avoid this by roughing the sides to allow water in.

Lay 2 inches or more of 1-2 inch diameter round river gravel in the ditch. (Crushed gravel takes a lot more energy, big machines and gas to make, and it's flat sides sit closer together, leaving less space for the water to flow in, but if that's all you can get, it'll do.) Next, lay the perforated 4 inch plastic pipe on top of the gravel. If you don't like plastic, you can go the old fashioned way and use cylinders of ceramic tile laid end-to-end. You will be creating an empty space for the water to flow along after it runs in through the perforations or between the tiles.

Tips for laying the pipe so water flows all the way

Start at the high point in the ditch and run the pipe or tiles at a slight slant to where you want the water to go. Here's a trick to make sure the pipe or tiles will drain the way you want them to. (You'll need at least a 1/4 inch drop for every 10 foot length.) Starting at the high point, lay a long, reasonably straight 2x4 on top of the pipe or tiles. Place a level on top of it. When the air bubble in the level is off center to the uphill side of the ditch, you know the water will flow. Hold the pipe or tile against the 2x4 and fill in underneath the pipe or tile with more gravel. Then move along the pipe or tiles until you get to the outlet. This ensures that you won't have low points where the water will pool instead of happily flowing down its new course.

When you have the pipe or tiles where you want them, fill the ditch almost to the top with round clean gravel. Fill it the rest of the way with 3 inches of straw and/or a 1/8 inch stack of newspapers to help filter out the dirt particles that might clog the spaces between the pieces of gravel, or block the holes in the pipe. (See illustration on page 55.)
Pathways or garden walls can be built right on top of the drains or cover the drain with topsoil and plants. Do not cover it with a clay soil because the water will have a hard time getting through it and into the drain.

THE FOUNDATION

MAKING THE FOUNDATION

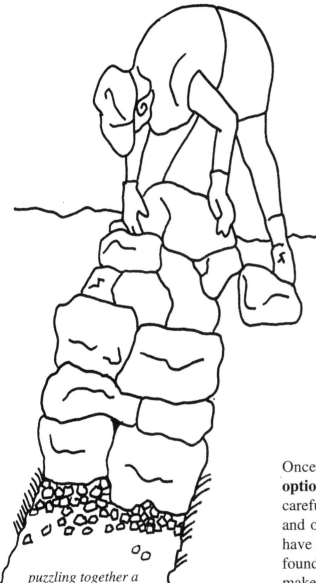

puzzling together a stone foundation

YOUR HOME IS ONLY AS STURDY AS YOUR FOUNDATION

Keep the purposes of the foundation in mind as you plan and build. The foundation goes under the walls:

- *to support the weight of the walls and roof (cob is very heavy)*
- *to create a stable base for your structure and to minimize the ground movement under your structure*
- *to keep cob walls away from ground moisture.*

Once **the foundation is made it will establish your options for all aspects of the building.** Read this book carefully before you start on your foundation. Libraries and other organizations in the natural building network have lots of information on how to build various types of foundations. Before deciding what kind of foundation to make, assess your options, materials, resources, and the time that you have available.

Give yourself lots of time for the foundation work!
You'll end up with a stronger foundation, and have much more fun making it, if you feel you have plenty of time to work on it.

PLANNING THE
FOUNDATION
(Footing or Plinth)

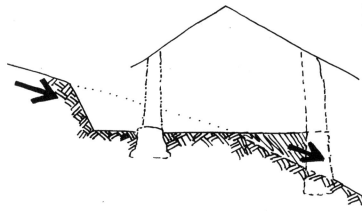

It's generally a good idea to **modify the land as little as possible.**

The soil is most likely to slip where it's been disturbed:
- where the bank has been cut,
- where soil has been added to make a level surface (see arrows).

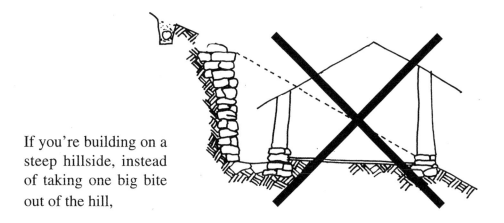

If you're building on a steep hillside, instead of taking one big bite out of the hill,

take two or more smaller bites, and make different floor heights.

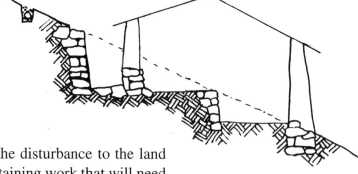

This will minimize the disturbance to the land and the amount of retaining work that will need to be done to discourage the soil from slipping.

You may decide to build up the floor on the downhill side of the house site.

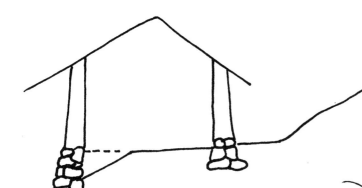

If you dare, and if the hill is not super wet, the retaining wall can second as a foundation. If you decide to do this, it will require lots of care to keep the water out of the house!

If you do, it's still a good idea to put the foundation on solid subsoil. If the hill is steep and you are adding a lot of in-fill to level the floor, build a heavy duty foundation that's tall enough to support the floor in-fill.

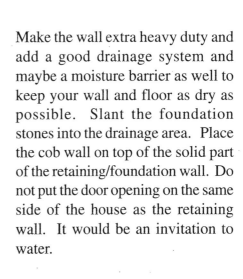

You might need to build a **retaining wall** on the uphill side of the house.

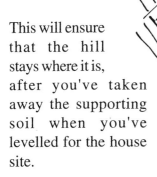

This will ensure that the hill stays where it is, after you've taken away the supporting soil when you've levelled for the house site.

Make the wall extra heavy duty and add a good drainage system and maybe a moisture barrier as well to keep your wall and floor as dry as possible. Slant the foundation stones into the drainage area. Place the cob wall on top of the solid part of the retaining/foundation wall. Do not put the door opening on the same side of the house as the retaining wall. It would be an invitation to water.

Designing the
door area

It is vital that you read the following chapter on floors (starting page 57) to help you design your foundation and threshold.

The positioning of the door(s) must be decided when you are designing your foundation. Underneath the door, the height of the foundation must be lower than the height of the rest of the foundation.
The depth of the foundation under the door may need to be increased so that there will be enough material to make sure the foundation is continuous. If you live where the ground is stable, you may opt to have no foundation under the doors or a much less substantial foundation under the doors and under any sunny, glass-filled, lightweight walls.

Decide what material you will use for the top surface of the threshold, and build it accordingly. If you are using the foundation stone itself, make sure it is as flat as possible so the bottom of the door seals well.

Make sure that the threshold will be a dry place.

* *Slant the outdoors ground away from the threshold.*

* *Plan to extend the roof eaves or add a porch to protect this area from rain.*

* *In temperate climates, place the door somewhere other than into the prevailing wind, or on the cold side of the house. In hot climates you may want to put it on the windy side of your house.*

* *Raise the wooden door frame above ground level to prevent rot.*

* *Where the ground level is a lot lower than the floor level, you may decide to make outdoor steps or a ramp up to the door. (During building you'll want a ramp for dragging tarps of cob and for wheelbarrows.)*

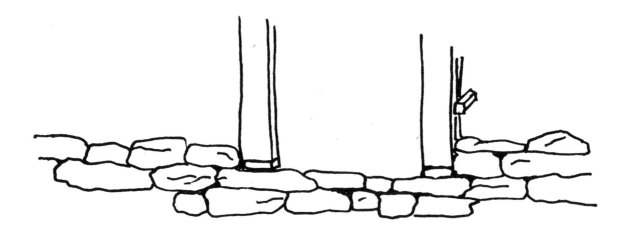

Consider **which way the door will swing** open. It invites you to go the way it swings. (If you live where it snows a lot, make the door open inward so you don't get trapped inside by snow piling up against the door.) However, doors opening in take up precious indoor space.

Setting up the door frame

Get your door and framing material. If you are not a carpenter, this is a great time to invite a carpenter friend over to give you a hand. When you've built the foundation to the bottom of the door opening, **set up the door frame, attach it to the foundation, level it, and brace it well.** Putting the door frame up as you build the foundation and before you start cobbing saves you future headaches. You'll be assured that the frame for the door will fit well against the foundation and the cob. Carefully build the foundation to the frame. You can fill in any gaps between the frame and the foundation with cob later. Add the keying system to the frame so that it will be attached to the cob as you build. (See page 111 for details about keying systems.)

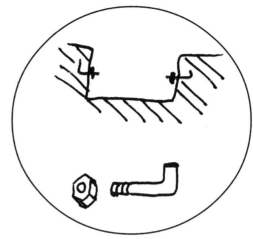

If you are doing a poured foundation, bury bolts into the concrete to attach the door frame to.

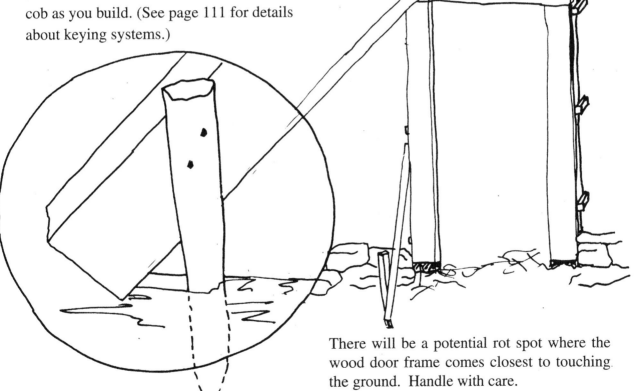

There will be a potential rot spot where the wood door frame comes closest to touching the ground. Handle with care.

Raise the frame off the ground with a flat stone or sturdy brick. Make sure that whatever you use will not get in the way when the door swings open.

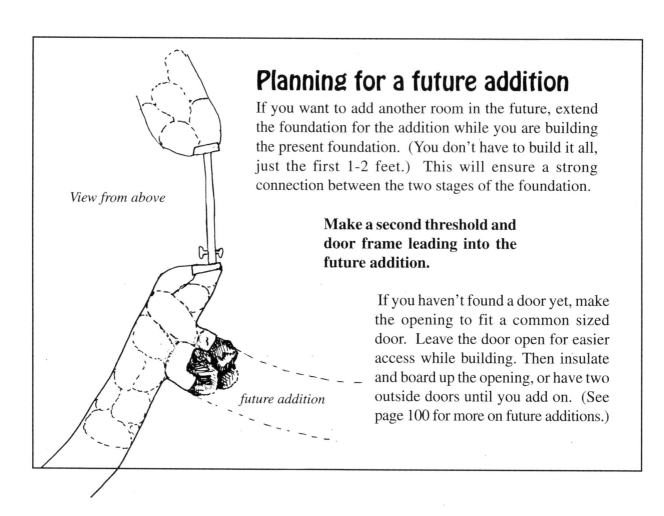

View from above

Planning for a future addition

If you want to add another room in the future, extend the foundation for the addition while you are building the present foundation. (You don't have to build it all, just the first 1-2 feet.) This will ensure a strong connection between the two stages of the foundation.

Make a second threshold and door frame leading into the future addition.

If you haven't found a door yet, make the opening to fit a common sized door. Leave the door open for easier access while building. Then insulate and board up the opening, or have two outside doors until you add on. (See page 100 for more on future additions.)

future addition

Other things to think about

- Where the wall makes a sharp curve, it is naturally strong. **The wall and the foundation at a curve can be slightly thinner (to save material and labor) than under the straighter parts of the wall.**

- Make the foundation and the wall extra-wide at door openings and in heavy support areas.

- If you plan on a very heavy roof, like ceramic or concrete tiles, add an extra inch or two of width in the supporting walls for good luck.

- The foundation under interior walls can be less substantial than for the outside walls, because there will be no need for protection from moisture or frost. Unless the **interior walls** will have the job of supporting the loft or roof weight, they **can be** thinner (**8 to 10 inches at the base, at least 5 inches at the top.**) Again, remember you can curve these walls for strength.

- Any **buttresses** you're planning will need foundations under them. (See design section, page 13 for more information about buttresses.) Built-in furniture that will be against the wall can have its foundations built into the main foundation at the same time, and can serve as little buttresses. Because built-in furniture will be supporting much less weight than a wall, it will need a less substantial foundation.

Tamping tips

Regardless of what type of foundation you decide on, always tamp the soil on the bottom of your foundation trench to compress the soil and minimize future movement.

A simple tamper can be made from a heavy piece of a tree (4 or 5 inches across). Either drill a hole and stick a dowel or pipe through the hole to make a handle, or chop a handgrip with a hatchet. Another option is to screw a piece of plywood (1 foot square + or -) onto the bottom of your tamper. The tampers can have bells or bottle caps nailed loosely to add some percussion to the tamping rhythm. (A 'tamporine'.) Another simple tamper can be made out of a 2 1/2 inch (plus or minus) metal pipe filled with packed dirt or stones and capped on the ends. You can use the special caps they make for the tops of fences. The metal pipe can have a flat piece of steel (1 foot square, + or -) welded onto the bottom. The smaller the bottom of the tamper, the more pounds per square inch you'll get out of it.

Tamping is hard work. Do a little at a time and pass it on to the next person. Remember to keep your knees bent, the tamper close to your body, and breathe a lot. To save effort, lift the tamper and let gravity do the tamping. Even though you might not be able to see the results, this is a very important step and will do a lot towards stabilizing the ground under the house.

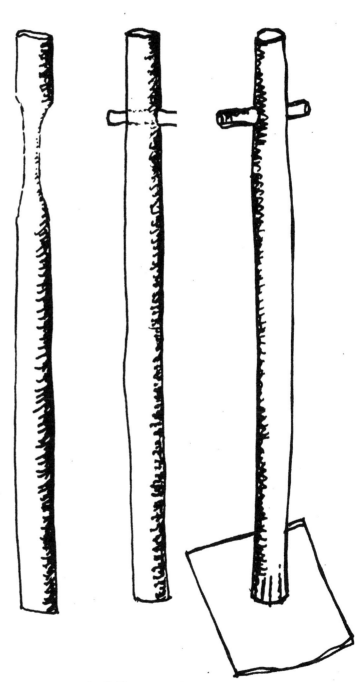

Always tamp the ground before building onto it.

How deep do I make the foundation?

Check with your local planning department or building contractors about foundation requirements and customs in your area. They will be able to give you information on the types of soil, how deep you'll have to go to avoid freezing, and the likelihood of earthquakes. Usually they go for overkill, but this will give you some useful information.

Scrape off the topsoil and put it on the garden area. **Dig down (at the very least 6 inches) to SOLID subsoil or rock.** You can tell when you get to the subsoil because it is so much harder to dig. Dig down to where you will be reasonably safe from frost heave. Pile the soil somewhere handy for making cob mixes later.

Roughly level the base of the foundation ditch.

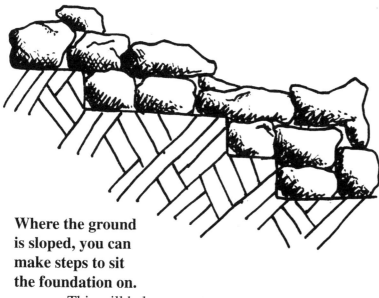

Where the ground is sloped, you can make steps to sit the foundation on.
This will help prevent the house from sliding down the hill. You can even angle the steps slightly into the hill.

Dig out any roots near the foundation. Live ones can grow into the foundation and pry it apart. Big dead roots under the foundation will decay, leaving a hollow spot in the ground under your house.

For those of you who live where the ground freezes deeply, a substantially deeper foundation will be needed. Doing a deep rock foundation will take a lot of rock, dedication, effort, and time. If you want the look of a stone foundation, you may want to make a reinforced poured-concrete base up to ground level for your stonework to sit on. Embed the first layer of stonework into the top of the wet concrete.

You may decide to make the foundation beneath the door deeper than the rest so you can create a strong continuous foundation. (See illustration page 36.)

Often a few inches of gravel are placed directly beneath the foundation. If you decide to do this, make sure you dig the trench that much deeper for the gravel. Tamp the ground before you put the gravel in, then tamp the gravel after it is in the trench.

The gravel under the foundation is good for:

- a secondary drain in case water makes it past your drainage system
- minimizing frost heave damage in very cold climates
- making a stronger base in very soft soils
- all the drainage you'll need in very dry areas.

How wide do I make the foundation?

To estimate how thick your cob walls are going to be, and therefore how wide to make the foundation, see the wall building section (page 86).

Make the ditch for the foundation as wide as the foundation will be, plus enough room to work in comfortably.

Where the ground touches the foundation, the cold or heat of the earth will be transferred into the foundation and into your dwelling. **Where temperatures are extreme, insulate the outside of the foundation with something that blocks the transfer of cold or heat** and does not absorb water. Modern construction folk use that blue Styrofoam stuff. In the United States, it's called 'Formular'. We've yet to come up with something natural that works well as an outside foundation insulation. Even gravel might help, though. (Refer to page 55 to see where insulation goes.) If you plan to insulate the foundation, leave room for that, too.

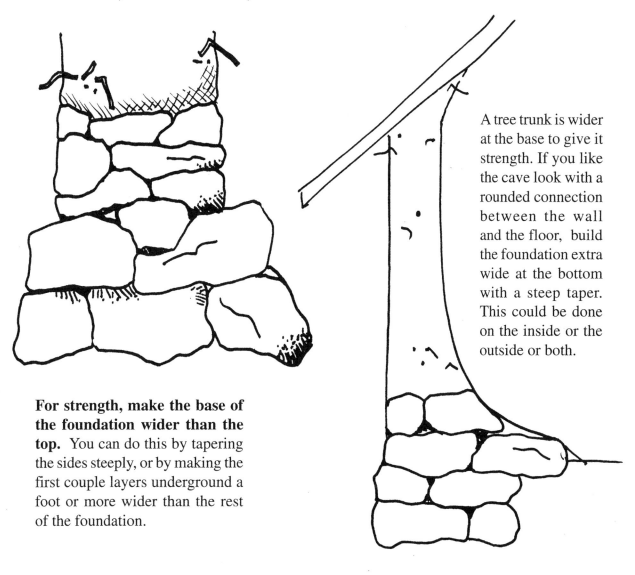

A tree trunk is wider at the base to give it strength. If you like the cave look with a rounded connection between the wall and the floor, build the foundation extra wide at the bottom with a steep taper. This could be done on the inside or the outside or both.

For strength, make the base of the foundation wider than the top. You can do this by tapering the sides steeply, or by making the first couple layers underground a foot or more wider than the rest of the foundation.

How high do I make the foundation?

Build your foundation at the very least 8 inches higher than ground level. Make it higher if you live where it's wet, where there's a lot of windblown rain, or if you simply love stone work. If you are making a two-story building, make the foundation at least 16 inches high. A common height for a foundation is 18 inches above ground level. You can check with the local builders to find out what they suggest in your area.

Moisture barrier between the foundation and the cob?

Mortar, concrete and earth slowly draw up moisture. One would think that the obvious solution to this would be to put something waterproof somewhere in the wall to block the moisture from the ground. Easier said than done! Wherever there is a moisture barrier and temperature differences, moisture condenses. Moisture barriers limit the ability of the wall to breathe. If moisture gets in, as it invariably does, a barrier will slow down the drying process.

I haven't noticed any moisture problems in the structures we've built with stone foundations, thanks to their good drainage systems.

Creating a good drainage system is the best way to keep your foundation, floor, and walls as dry as possible.

Another minus to using a moisture barrier is that it can weaken the connection between the cob and the foundation. There are two schools of thought in the natural building fields on how to think about moisture barriers.

Some say the breathability is vital, never use a moisture barrier. Others say they work OK on top of the foundation to keep the moisture from wicking up into the wall.

The most natural type of barrier I've heard of is flat stones embedded on the top of the concrete or soil cement foundation to stop the upward flow of water from getting into the wall. Remember to slant the stones toward the center of the wall and towards each other so the cob doesn't slide off the wall. (See illustrations on page 49.)

Getting plumbing and electric wire into the house

Plumbing and electricity are pretty easy to do yourself. The library or a knowledgeable friend is all you need to get an understanding.

If there's any chance you'll want electricity and/or indoor plumbing, be sure to lay some 4 inch (or bigger) pipes through the width of the wall or foundation for the wires and the water pipes to run through.

If you bury the pipes between the foundation stones, make sure the stones are not leaning on the pipe but onto other stones. If you live where it freezes, the water pipes will need protection from the frost where they are exposed to the outside air. Angle the pipes for the plumbing slightly down towards the outside, so that if a leak develops, it'll run out. (See wiring section on page 99.)

You'll need a water intake <u>and</u> a drain outlet. These can be put at the bottom of the foundation (for buried wire and pipes) or through the foundation.

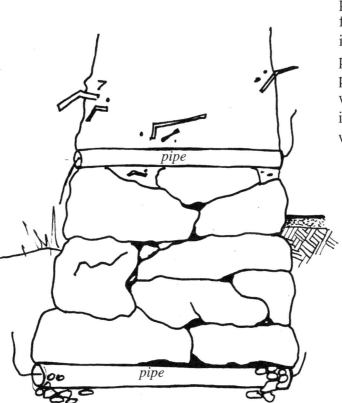

You can put the access pipes on top of the foundation and cob them in place. The wiring and plumbing is usually pretty ugly so you may want to put them lower in the foundation so they will be less obvious.

Any extra space between the access pipe and the plumbing pipe or electric wire can be filled later with cob or plaster.

Firewood box in the foundation
Some folks put a firewood box opening through the foundation. The box can be filled from the outside so you won't have to carry firewood in through the house. Make the foundation extra deep if you want the foundation to be continuous.

Stone Foundations

Stone looks the best and will still look the best a few thousand years from now, after your home has returned to the elements. The easiest foundation I've ever seen was a giant boulder with the house built right on top of it! Unless you have such a thing, you'll be gathering smaller stones as pieces for your foundation puzzle.

Choosing stones

Ideally, you'll be able to gather stone from the site or nearby. Old quarries and along nearby road cuts are good places to scrounge rock. When you start collecting, you'll see beautiful ones all over! A 20 foot round building will require roughly 8 tons of stone. Be conscious! Remember, gathering materials impacts the environment.

Find stones that have **two somewhat parallel sides.** The more brick-shaped your stones are, the easier they are to build with. Make sure to pick ones that are up for the job. Leave soft, cracked, flaky ones where they are. When in doubt, throw them down on the ground to see if they'll break.

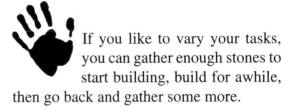 Sometimes you may want to modify the shape of a stone. A steel mallet can be used to whack off unwanted protrusions. Practice helps you learn how rock breaks. Of course, each one has its own personality. You'll get a feel for it. **Wear eye protection.**

If rocks are not available, big broken-up concrete pieces make a good substitute. They stack easily because they usually have two parallel sides. Sometimes construction companies will even deliver their garbage concrete to your place. If the pieces are too big to handle, whack them with a sledge hammer. **Remember to wear eye and ear protection.**

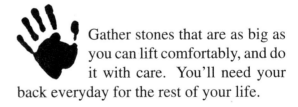 If you like to vary your tasks, you can gather enough stones to start building, build for awhile, then go back and gather some more.

Gather stones that are as big as you can lift comfortably, and do it with care. You'll need your back everyday for the rest of your life.

If you would rather leave the collecting job to someone else, cement suppliers and landscape firms often have rocks for sale by the truckload. Check to see that they're suitable for building before you buy. They will deliver them.

Suggestions for moving rocks

Get one of the dumping-type wheelbarrows with two wheels for transporting the rocks. These are sometimes called gardening carts. This will save a lot of work. I think they're wonderful.

 With one of these you can simply:

1)

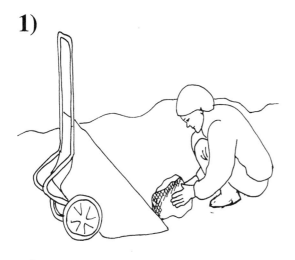

roll the rocks in,

2)

lever the barrow to the upright position

3)

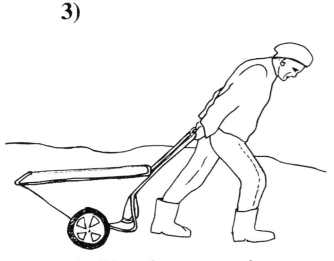

and <u>pull</u> it to where you want them,

4)

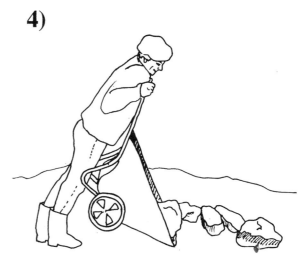

then dump.

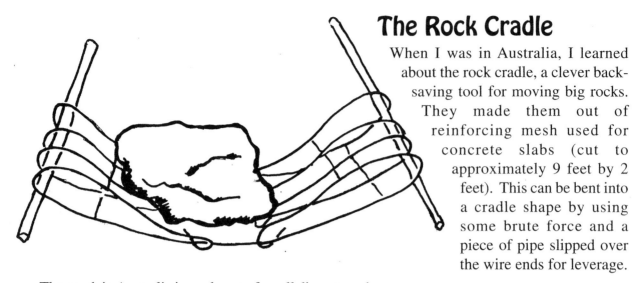

The Rock Cradle

When I was in Australia, I learned about the rock cradle, a clever back-saving tool for moving big rocks. They made them out of reinforcing mesh used for concrete slabs (cut to approximately 9 feet by 2 feet). This can be bent into a cradle shape by using some brute force and a piece of pipe slipped over the wire ends for leverage.

The mesh in Australia is made out of small diameter rebar and is much more heavy duty than the mesh I've been able to find in my town in Oregon. I tried making one out of the reinforcing mesh that I could get, and it wasn't up to the job. A welding friend may be able to help you create a strong enough mesh for a rock cradle. It may be possible to adapt this design and make a rock cradle out of heavy duty canvas.

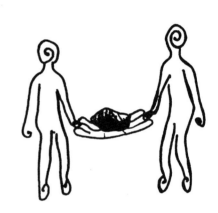

A large stone can be rolled onto the cradle and one person on each side of the cradle can safely share the weight of the stone. You'll need to wear gloves so the metal doesn't chafe the skin on your hands. For bigger stones, a strong stick or crowbar can be stuck through the handles of the cradle and four or more people can carry the weight.

THE ROCK GAME

The rock game is a living room/lounge game that will teach you a lot about puzzling rocks together for your foundation. It's a very entertaining right brain game for all and a great conversation piece!!

Collect special pretty rocks of different shapes, sizes and colors. Get some wedge shaped ones and some concave ones. Too many flat ones are boring. Keep your rocks for the game in a big bowl or large wooden tray. You can use a special little rock game rug to protect your floor.

One person decides which kind of structure to make in the beginning of each game:

multiple arches, domes, little foundations, towers, or getting all the rocks on. Go around the circle taking turns adding a rock. Make up your own rules as you play.

You can get new rocks and throw out 'old' ones anytime as you learn which kinds of rocks stack well and which ones don't. This game will help you learn which rocks to collect for your foundation and how they fit together.

I hope you enjoy this game as much as I do!

Thanks Robbi and Ashley for teaching me this wonderful game!

Making a stone foundation

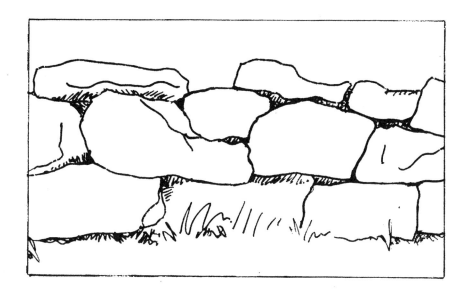

Leave plenty of time! Building a foundation is a delightful puzzle. Well, that depends on what kind of person you are. As far as I can tell there are two kinds: the ones who love stone puzzles and the ones who hate them. I've noticed that some days I'm on and others I'm not, so give it more than one day to decide which kind you are. If you fall under the second category you might choose to do some other kind of foundation, or find a friend or two who love stone puzzles.

It makes sense to spend a lot of time looking at the rocks to see what shape you need and which rock fits where, instead of using your muscle to lift and fit them together by trial and error. Take your time. Hurrying and rock work are opposite concepts. Rocks have been around for a long time and they like to move slowly. Stonework at its best is an incredible meditation.

Some people say to wear sturdy shoes and gloves for stonework. That won't help nearly as much as keeping your soft little fingers and toes out from between a rock and a hard place. If you're working with someone else, be careful of each other's fingers and toes, or work on separate sections of the wall. Be mindful!

Remember to tamp!

The bottom is a good place for the biggest, heaviest stones. Find where they want to sit firmly and snuggle them up close to each other. Keep in mind that you are also creating the base for the next layer of stones to rest on. Remember to **make the bottom of the foundation wider than the top.** Moving stones around at or below ground level is especially hard on the back, so take it easy. Lay stones all along the bottom of the foundation trench. (See page 50 to help you decide whether or not to use mortar.)

47

Cover each vertical crack of the preceding layer with a stone.

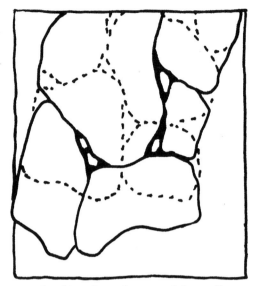

Looking from the top of the wall

Stones have a mind of their own. They seem to know where they want to be. Don't argue with them: you will lose every time. Make sure each stone is happy by standing on it and rocking it. (Is this where the word "rocking" comes from?) If it wiggles, move it until it's secure, or use a small strong rock as a wedge to stabilize it. It is easier to stabilize a stone that is resting on three points than on four. Long stones that run the whole width of the wall every once in a while add strength. Continue along, carefully fitting each stone into its place. **Fill any empty spaces with small stones:** they help hold their big sisters in place.

Patience and practice are very useful at this point. Remember to stand up often and stretch your back by leaning backwards to re-squish the discs between the vertebrae back into a rounder shape.

Drink lots of water! Too much rock work at a time is too much rock work. Take it easy. Warning! For the folks who love puzzles this can be very addictive.

As soon as you get **above ground level, the shape of the foundation will be permanently visible.** Stand back and look at your artwork from a distance. Do you want an interior wall that is perfectly vertical all the way down to the floor? If so, build the foundation so it will line up with the vertical plane of the wall. On the outside, do you want the same angle as the exterior wall taper (2 inches in 3 feet) or do you want the super stable look of a steeply tapered base? Build accordingly.

If you want to see exposed stone, use ones that create the look that you want to see on the sides. Save the prettiest stones that you want to see forever and use them above ground near the entrance or patio areas. To highlight these you can stick them out a little from the plane of the rest of the wall. Use your artist's license.

Continue building until you reach the height you've decided on. The top needs to be rough so the cob has something to hold on to. It's best if the **upper facets of the top layer of stone angle either into the center of the wall or into each other.** If you are using mortar, do not make any places on the top where water will pool.

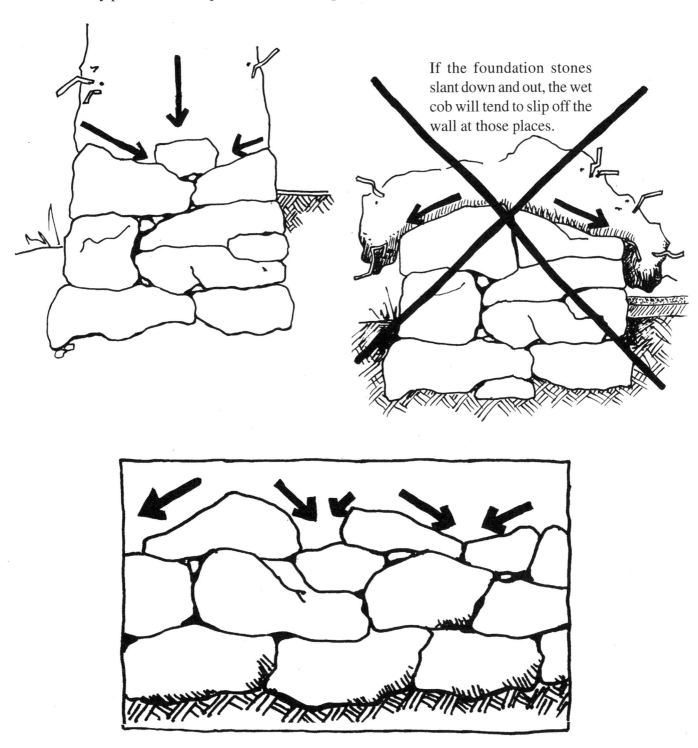

If the foundation stones slant down and out, the wet cob will tend to slip off the wall at those places.

To Mortar or not to Mortar?

The purpose of mortar is to discourage the stones from moving.

Options
No mortar

If you have good stones and you're good at fitting them together, you can use no mortar at all. Many of the old cob homes in England are built on dry stacked stone foundations. You can cob in the gaps between stones from the outside to keep wind and critters out.

Cob mortar

Many of the ancient Anasazi buildings (in the Southwestern United States) were built using cob or earthen mortar, along with a lot of attention and skill in placing the stones. Some of these earthen mortars have a little lime added to them.

Lime mortar

Some ancient sites and many European walls were built using lime mortar between the stones. This was basically lime mixed with sand. You'll have to do some research to find out more about lime mortars. I have never used them so I can't give you much help.

Concrete mortar

This is the modern version of earth or lime based mortars. It consists of sand, lime, cement and water.

Concrete and rebar wall with a stone facade

This is what stone walls have devolved to. The library will have books on how to make these modern fake-stone walls.

to mix up concrete mortar

The recipe is as follows: **one part premixed *mortar cement* to three or four parts sand** of various sized particles, and water. (You can make your own *mortar cement* by using 2/3 portland cement with 1/3 dry builder's lime instead of premix.) Use three parts sand if you follow instructions well, four if you are conservative and want to stretch the cement a little farther. The production of cement is very bad for the environment so the less you use, the better. **Caution: cement burns skin and eyes! Wear gloves and eye protection.**

When you make up the mortar, mix small batches and use immediately. First mix the dry ingredients well, then add the water. It will start to cure as soon as the cement gets wet. Mix it in a wheelbarrow using a hoe or trowel, or on a piece of plastic or tarp by shifting the stuff back and forth.

You can use a wet mortar, or a dry one. When making a wet mortar, watch out because it will go from too dry to too wet very quickly. Add the water carefully until the mix makes stiff peaks. If it does get too wet, it's OK to add a little sand to stiffen it up. When using a dry mortar, barely moisten your ingredients. Mix well. Then it's ready to use!

tools you'll need for mortar

--trowel
--wheelbarrow, cart or plastic sheet for mixing
--gloves
--water
--hose and spray nozzle, good for lightly wetting the curing mortar
--hoe for stirring, there are special hoes for this job with holes in them
--whisk broom
--wire brush
--tarp or something to cover the curing mortar

mortaring the stones

Squish mortar into the cracks between the stones you've already set, wiggling it in with a trowel or stick until the holes are full. You can add small stones too. Do this after each course of stone is laid. **Clean your tools, gloves, and wheelbarrow before the mortar hardens on them.**

For a dry mix, barely moisten the ingredients and wet the surfaces of the stones generously before troweling the mortar into place. If you are using a dry mix for your foundation, you may decide to make up a wet mortar for the places between the stones on the very outside surface of the foundation - the dry mix tends to fall off the sides of the foundation before it cures.

curing the mortar

Cover the fresh mortar with a tarp or with damp burlap bags.
If you've laid the mortar in the morning, you can come back in the evening or the next day and tidy up any joints and stone faces that will be visible. A wet whisk broom will do the job unless the mortar has hardened too much, in which case use a wire brush.
Keep your mortar work covered for a week or so. Spray water on it every day so it doesn't dry out too fast. This is very important to the strength of the mortar.
When you add more stone to a partially cured section of the foundation, be gentle so you won't dislodge yesterday's work.

Some Other Foundation Options

Regardless of which type of foundation you choose, remember to tamp the ground under it and put in the pipes for the plumbing and electricity.

Poured concrete

Think twice about using a lot of cement! Its production requires mining, transportation, and heating materials to very high temperatures twice. This entails burning fossil fuels and often creates hazardous waste. Concrete makes an enormous contribution towards ozone depletion and air pollution.

It's common to add reinforcing steel to a concrete foundation. The library will have lots of information about how to reinforce this type of foundation.

Remember to add steel anchor bolts to attach the door frame to the foundation. Be aware that wood meeting concrete forms a potential rot spot. Wood treated with poison will fight off the rot longer than untreated wood. Some builders also put a moisture barrier between the concrete and the wood.

A **standard concrete mix** is made of 1 part portland cement, 2 parts sand and 3 parts gravel. You can mix it yourself or get it delivered already mixed to the site. Wet concrete does not stick well to dry concrete so ideally it should be poured all at once. Embed sharp-edged stones on the top before it cures. This will make a rough surface for the cob to hold on to.

Make sure the concrete cures slowly by keeping it wet and covered for a week or two before starting to cob.

Soil cement

Soil cement is otherwise known as 'poor people's concrete'. Soil cement is weaker than concrete but it is a good option for areas with limited rock and no road access, or for people who want to minimize their use of cement.

Depending on your soil (sandy is best) add up to 9 parts soil to 1 part cement plus water. Stir well. Use immediately. Make up some test batches with different amounts of cement and from different soil types until you find a suitable mix. Remember to put some stones on top of the soil cement for the cob to hold onto.

Soil cement can also be used for floors or made into large flat bricks for stepping stones or floor 'tiles'.

forms for poured concrete or soil cement

Any poured foundation will need forms to hold the cement mixture in place until it's cured enough to hold itself up. The library will have information on building forms and pouring concrete.

Conventional forms are made out of wood. If you'll be building a rounded house, you'll need a rounded foundation which is very difficult to form using wood. Scrap sheet metal held in place with stakes driven into the ground, or straw bales with stakes holding them in place, are simple, inexpensive forms for curved foundation walls. Tie the tops of the stakes to each other to ensure that the concrete doesn't push the forms apart.

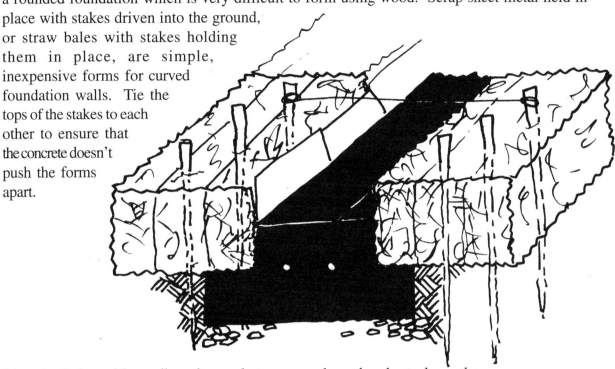

Line the bales with cardboard, or whatever you have handy, to keep the concrete off them. The straw bales will be used later in the cob walls.

To curve a straw bale, lean it up against something at a 45 degree angle and apply weight to the middle of the bale until you have the shape you want.

Concrete blocks mortared together

The library will have books that can tell you about this common way of building a foundation.

Railroad ties and gravel

As with any type of foundation, remember to tamp the gravel well and to make an adequate drainage system around the house. The creosote used in railroad ties is toxic, so you may want to use these for outdoor walls, sheds and furniture.

Earth-filled tires

You may want to set these on a tamped, gravel-filled trench to prevent possible rising damp. Pound earth into the tires with a big sledge hammer. Stagger the tires on top of each other like you would bricks. For more information on this, look up "Earth Ships" at the library.

Agricultural bags filled with earth and tamped

The white woven plastic seed or fertilizer bags are basically forms for rammed earth bricks.

If you're building where it's very dry, this type of foundation can be laid right on the tamped ground. If not, it's best used as an above ground addition to a stone or concrete foundation. Or build it on top of a gravel filled trench, because the earth in the bags may wick moisture.

Bags are filled while in place with slightly dampened, sandy earth. The tops of the bags are secured by folding them over and pinning them with a nail, or folding and leaning them against the last filled bag. Then tamp them well. Pound sharpened sticks or bamboo into the bags, so that the sticks protrude out for the cob to hold onto. The bags will disintegrate quickly in the sun, so cover them with plaster as soon as possible. You'll be surprised how well plaster will stick to the woven plastic!

I have heard that you can buy rolls of continuous bag material before it's been cut up for bags to use for this type of construction.

Building with these and with agricultural bags are systems developed by Nader Khalili. For more information on this contact: 0376 Shangri-La Avenue, Hesperia, CA, USA 92345.

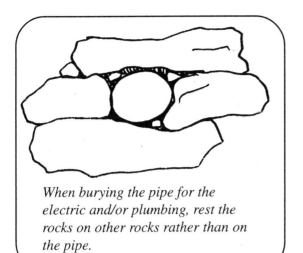

When burying the pipe for the electric and/or plumbing, rest the rocks on other rocks rather than on the pipe.

Foundation and drainage summary

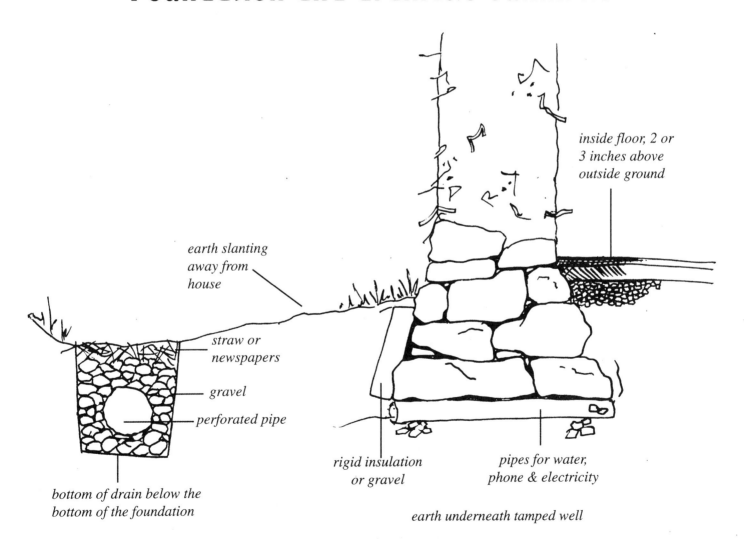

earth slanting away from house

straw or newspapers

gravel

perforated pipe

bottom of drain below the bottom of the foundation

inside floor, 2 or 3 inches above outside ground

rigid insulation or gravel

pipes for water, phone & electricity

earth underneath tamped well

55

FLOORS

Thank you Athena, Bill, and Robert for all you've learned and taught about earth floors! Books on patio floors might give you some other clever floor ideas. Remember, you can use more than one type of floor in different sections of your home.

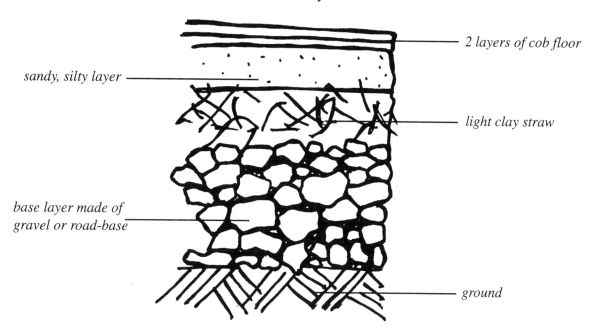

sandy, silty layer

2 layers of cob floor

light clay straw

base layer made of gravel or road-base

ground

General Info to Consider

Accessing the underground temperature

The **temperature in the ground below frost line is the average yearly air temperature.** (The temperature in the ground is the same all year round.) Find out what that temperature is and if it's a comfortable temperature you'll want to **connect your floor to the ground.** If your average yearly temperature is not pleasant, you may want to move to a more reasonable climate or consider insulating your floor from the ground. To get an idea, call the local concrete folks and ask them what the insulation requirements are in your area.

If some insulation is added under the floor, the heat absorbed by the floor will reflect back to you quicker, but you then lose some of your connection to that average yearly temperature. There may be data on optimum thickness for an insulated thermal mass floor in your climate.

When you build, you will be creating a protected zone in the ground under the structure. It will be shaded from the heat and protected from the freezing by the house. The indoor heating and cooling and the sun shining through the windows will also affect the temperature in this area. **If you live in a climate with extreme temperatures, you may want to insulate around the perimeter of the foundation down to the frost line** to keep the heat from

escaping through the frozen and/or wet ground surface. This is usually done by burying a special water resistant styrofoam (sometimes called Formula) between the foundation and the ground all the way around the house. (2 inches is rated at R10, 3 inches at R15.) The styrofoam stuff needs to be protected from the ultraviolet sunlight. Check with local builders or the building department about how deep, and how much to insulate in your climate's conditions.

Collecting passive solar heat with the floor

The floor is one of the best areas to store the sun's heat, because it's where most of the sun rays hit after coming through the windows. This is why it is valuable to **make the floor out of something that has plenty of thermal mass.** That means make the floor out of something heavy and dense like earth, brick, or stone.

When I lived in an old house with an uninsulated concrete slab in a temperate climate, I was amazed at how comfortable it was. In the hot summers, that house was cooler than all my friends' houses. And in the winter, the floor was cold to the touch but that house needed a lot less heating than I expected. I put down rugs in the areas where my feet touched the floor and was cozy enough. In the rooms where the sun shone onto the floors, I left the floors bare so they could soak up the heat from the sun.

Cold sink

Cold air is like water. It will flow to the lowest point on your floor. If you design your floor with a low spot, the coldest air will flow off the floor where your feet hang out and into the lower cold air reservoir. Put the cold, sink spot on the cold side of the

house. Because this will be the coldest place in your home, it is an ideal place for cold food.storage if you don't have electricity, or if you'd rather not have a refrigerator. You might be delightfully surprised, with a little adjustment to your habits, how easy it is to live without an electric fridge. If you live where it's very hot, you may want to make the cold sink area a living space, a cool bedroom, or summertime living room.

Stepping the floors to different levels is an easy way to deal with a sloped site. The differing floor levels define and separate indoor spaces and enhance the interest in a room. Steps from one floor level to another must be at least 4 inches high and at most 9 inches to prevent stubbed toes. The step(s) where the floor level changes can be flowing organic shapes or straight edges. Take this into consideration when planning your foundation. **During construction, make temporary ramps where the steps will be.** This will make life easier while you are dragging the tarps full of cob around the floor area during cobbing. Make the actual step later when you've finished the walls.

Planning the floor at the door threshold

The door's threshold limits your choices about the height of the floor at the entrance. When designing the foundation and threshold, it is helpful to estimate floor height(s). **Ideally, the threshold will be at the same height as the floor surface.** Your **floor can be built directly on the ground.** Read on to help you decide how thick your floor will be.

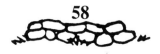

Make the floor higher than the ground outside

Avoid 'CSP Syndrome'
(covered swimming pool syndrome).

Plan for the finished floor to be **at least 2 inches above the outside ground level.**
Remember, the ground outside can be lowered as you dig soil for constructing the wall. Make sure the ground slopes away from your home in all directions.

Making the floor

Tamping the ground

It's easiest to decide the floor height(s) early on in the design and construction processes. That way you can do the rough leveling of the floor area before you start to build. It's also best to decide what kind of floor you are going to make before you start building, so you can dig down to, or fill in to, the appropriate height. If you leave it until after the builders have tramped back and forth across the ground, it will be a lot harder to dig the dirt out to level it.

Scrape off the topsoil in the floor area and put it where your garden will be. Any subsoil that you dig out of the floor area can be used for making cob. The building process and all those feet will help do a lot of the tamping work for your future floor. If you are building up the floor level, you can do that before you build or anytime during the building process to make the most of the tamping feet of the builders.

Go over the ground and each layer of floor base with your tamper. You can't tamp too much. You may even want to rent or borrow an electric powered vibrating compacting machine to save you work.

Finding level for the floor

To help you estimate a level floor surface, use a long level, string level, or water level. If you use a long level, sit it on top of a straight 2x4, like you used for getting the slope for the drainage. (See the illustration on page 31.)

using and designing a water level

Attach a 3/4 inch clear plastic hose to a container, a bucket, or an old kettle. Put a clamp or a tap on the hose so you can open and close the flow from the container to the hose. Set the container in a secure spot where it's above the highest point on the floor area. Fill the container with colored water. Hold the free end of the hose above the container and open the clamp so the colored water can flow into the hose. The water in the hose will always seek the same level as in the container, so you can establish even floor heights by measuring down from the water level in the hose all around the floor area. There are other types of water levels. Any sort will do the job.

marking your estimated floor height

Knowing the floor height will help you measure the heights of the counters, seats, windows, etc. as you build. Mark the future floor height on stakes pounded into the ground inside the floor area. If you plan more than one floor height, mark each one with stakes. Put the stakes somewhere out of the way because they can be perfect toe stubbers! Once the foundation is made, the floor height can be marked along the inside of the foundation with a chalk line or pencil. Give yourself plenty of space on either side of a door before raising or lowering the floor level.

Base (or layers of base materials) under the floor

(See illustration of layers that make up the floor, on page 57).

Reasons for putting a base under the floor:

- Usually the **floor needs to be filled in so that it is at least 2 or 3 inches higher than the outdoor ground level to insure it will stay dry.** You may have to bring in some fill to raise the floor. **It's OK if the floor's base makes the floor higher than the foundation, and makes the floor butt up against the cob walls as long as you've planned for this when placing the doors, and heights of the seats and counters.**

- **The base helps the floor to breathe, and therefore keeps it drier.** The air spaces between the particles in the base discourage capillary action of water that might sneak in under your floor area.

- **The base is made of stabilizing materials that flex, minimizing the stress on the floor when the ground moves.**

- **The base provides a little insulation which speeds up the time it takes the heat to be reflected back into the house.**

Moisture barrier between the ground and the floor?

The general rule with natural building is to avoid moisture barriers if possible. With good drainage design and an adequate gravel base, your floor shouldn't need a moisture barrier.

Floor base tests

You can make floor base experiments in 4 or 5 inch plastic potted plant containers. Fill them with the materials you have available, and tamp each layer. Then push them out of the container molds. See which are the toughest. **Regardless which base you decide on, tamp, tamp, tamp!**

Floor base options

the ground as a base

cob floor layers

ground

It is possible that the ground is OK just as it is. A good drainage system around the house should keep your floor dry. Some soils are very stable and drain well and an added base may be unnecessary. Remember that the finished floor needs to be higher than the outside ground level.

gravel as a base

There's this great stuff you can get at the concrete supply place called road base. It's what's used under the roads for drainage and stabilizing. It is crushed rock in various sized pieces, usually 3/4 inch on down to powder (called "3/4 minus" in America).

Four inches (or more) of road base, dampened and tamped, make a very stable and very cheap base for your floor. Some people sift it and put down layers of the different sized particles that separated out, starting with the largest and layering up to the smallest sized particles.

You can use **gravel (river or crushed) or shale for the base** of your floor instead.

Remember to tamp it well. If you live where the ground water level is high, you may want to make an extra thick (10 inches or so) base or a layer of large gravel and a layer of road base. The gravel will discourage water from wicking up into the floor. **If you use a thick base for your floor, it is important to make your ceilings that much higher too.**

sandy or silty soil layer as a base

A leveled layer of **dampened, tamped,** sandy or silty soil, two or three inches thick, is sometimes used as a base or as an added layer on top of the gravel or road base layer. This needs to have enough clay in it to bind itself together when it's dry. If you are making a floor of flat stones and bricks, the sandy or silty soil makes a good bedding for them.

You can put **pipes under the floor to heat your house.** Hot water circulates through the coils of pipe, heating up the floor and the interior space. I have never done a floor with heating coils in it, but there is information available on putting heating pipes into concrete slabs. You can modify this information using natural materials. If you want to bury pipes under the floor, you can put them in this soil layer, but only if you are not adding a screen layer that will insulate the heating pipes from the interior.

optional screen layer

Some natural builders put a 3 or 4 inch layer of tamped light clay/straw (see page 143 for instructions on making light clay/straw) over the gravel before putting down the soil. This **prevents the soil from sifting into the spaces between the chunks of gravel and adds a little layer of insulation.** The light clay/straw is tamped well or compressed when you tamp the soil layer. It will crush down to an inch or less. Loose straw or old sheets or newspaper would serve the same screening purpose, like on top of the drainage ditch. If you are troweling a cob floor right onto the base, you won't need to worry about it filling in the air spaces between the gravel.

Other possibilities for an insulating base layer are to mix clay slip with vermiculite, perlite, or pumice.

Cob floor surface

A one-piece cob floor is the longest lasting type of earth floor. The floor surface sits on top of whatever you've decided to use for the base layer.

*You can make a floor with **one 3/4 inch layer or two 1/2 to 3/4 inch layers of your cob floor mix** (with optional additives) thinned down to a consistency that can be troweled on. Make it as dry as possible but wet enough to trowel. The less moisture there is when you lay it down, the quicker it will dry and the less it will crack. Keeping your trowel wet while working on the floor will make it a lot easier to smooth the surface.*

tools you'll need

- shovel
- wheelbarrow, cart, or tarp for mixing
- a 4 foot level taped to a 2x4 on edge and/ or a water level
- chalk line
- pencil
- trowels
- bucket for storing some of your dry mix for future repairs
- plywood squares for standing on to protect the floor surface

Floor surface recipe

The floor is basically the same stuff as the walls, sometimes with extra sand added depending on your original mix. (See page 72 for proportions for making cob.) Sift all the ingredients quite finely for the final layer of floor.

If you use straw in your floor mix, chop or grate it finely. **Use shorter pieces of straw and less** if you don't want it to show in the floor surface. The straw can be grated quite small. (See page 153 for information on how to break up the straw.) If you have a clay rich soil and a shortage of sand, use more straw instead of the sand. Some natural builders don't use any straw in their floors. Make tests until you find a recipe that works.

Possible additions to a standard cob mix for floors

Go over the section on plaster additions (pages 154-156) and apply that info to the floor mix.

glue. Adding a little Elmers glue helps harden the floor.

ground psyllium seed husk. This is a wonderful addition to a cob floor! Have you ever noticed tennis courts or tracks that feel like "soft" concrete? Psyllium gives them that soft, rubbery feel. This makes the floor easier on your legs and on your dishes. A little bit of this stuff goes a long way. Too much will make it hard to trowel. Make up some test batches with different amounts of this additive in each until you find the percentage that works best. For a 12 foot diameter round room, you'll need about a pound of psyllium.

manure. A very common ingredient for earthen floors in many parts of the world. This can be used as the fiber in your recipe.

blood. Ox blood is a common ingredient in old floor recipes. It's supposed to make the floor a lot harder. These days you can buy blood meal at plant nurseries. Use this on the top layer.

wood ash. This is an additive that I have come across in old literature. I haven't tried it yet.

oil. A little bit of oil can be added to the last layer of floor. You'll probably want to use an oil coating on the finished floor, in which case adding oil to the wet floor mix would be unnecessary.

flour. Like with plaster, the addition of flour paste will harden the surface.

test batches for the floor surface

Try out different floor recipes before you decide which one(s) to use. Do lots of tests. Outdoors, make little samples of the base material and trowel the test mixes onto it. A two foot round sample will be plenty. Let the samples dry, protecting them from direct sun and from rain. How do they look? See how they hold up when you walk on them. Once you've narrowed down what works, make some larger test batches of the same recipe to make sure it's the one you want. You can try out your homemade sealants (see page 66) on the samples too.

Make test batches varying the ratio of sand to clay, until you come up with a recipe that doesn't crack and holds up when you walk on it. Cracking means too much clay, and too much sand will cause the floor to slough off when walked on. If you are doing more than one layer, you can make the bottom layer 75-85% sand and small gravel.

Putting down a cob floor

Ideally, you'll be able to trowel down **each layer of the whole floor in one day.** You can mix up enough material for the whole floor layer a day or two ahead of time and let it cure. For a big floor, start in the morning so you'll have the whole day to put it down if you need it.

Make sure the base for your cob floor is as clean as possible. **The base then needs to be dampened.** Otherwise, the dry material under the floor will suck the moisture out of the floor quickly, and it will be more likely to crack. Spray it lightly again and again so it soaks in well but doesn't create puddles.

Start laying the floor furthest from the door, working your way toward the door. That way you can get out without walking over your beautifully troweled floor. Oops!

Working back and forth across the floor, **add a 1/2 to 1 inch layer of floor mixture in reachable swathes (2 feet or so).** Leave the edge slanting toward you (45 degreesish) and scratched to make a rough surface for the next strip of fresh floor mixture to adhere to. Spray the slanting edge with water when you're ready to attach the next bit to it. Lay your next strip of floor, etc., etc., until you're done!

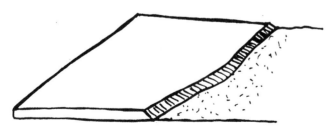

Around the edges, where the floor meets the foundation or the walls, is a tricky place. The already dry walls quickly sap the moisture from the floor where wall and floor touch.

Sometimes the floor will shrink away from the walls as it dries, leaving a gap there. You can take some of your floor mixture and let it dry a little, or mix in some dry ingredients until it's as dry as it can be and still be malleable. Squish this together in your hands, roll it into a sausage shape, and put it down around the edges of the floor. Trowel it flat as you do the rest of the floor. The drier mix will minimize the shrinkage. (Thanks Athena! Good idea!)

After the **first layer of floor has hardened to the leather-hard stage, you can add the next layer.** When your last layer of floor has hardened to the leather-hard stage, **go back over it with a squirt bottle and trowel if you want a glossy smooth finish.** Stand and work from your plywood squares to minimize footprints.

If you've used too much clay and the top layer cracks, you can add another very thin layer of floor. Remember to dampen what's there before applying the next layer.

How to connect one day's floor work to the next
Sometimes it turns out that you can't do the floor in one fell swoop, or that you need to join the floor of a new addition to an old floor. Where the floors connect is a potential weak spot. If you know you'll be adding on to a floor, make sure you leave the edge sloped and scored well. That means scratch it up so the surface is rough and the next layer will be able to key into the scratches. When it's time to add on, wet the sloped edge really well and trowel down the new floor.

Save some of your original dry mix so you can use it for repairs later on.

Leveling the floor

The floor doesn't have to be exactly level. Slight undulations make a floor feel more natural and friendly. Have you ever seen anything (other than water) perfectly level in nature?

Establish where you want the floor height, and draw or snap chalk lines on the walls or foundation to guide you when you're troweling down the floor.

You can use a 2x4 with a level taped to it, and drag it across the wet floor to "scrape" it level. You can lay down guides running parallel to each other and close enough for the 2x4 leveler to span. These can be made of thin strips of the floor mixture or thin strips of wood that need to be removed as soon as possible and the gap they made filled and troweled.

Here's a little trick to prevent you from getting too far off level. While you are putting the base in and/or troweling the floor surface, nail big headed nails into the ground with the tops a tiny bit lower than the final floor height. Start with one nail and level the tops of the other nails to it and then to each other. Put them as far apart as your longest level will reach. This will give you a guide while you are putting down the floor. The last layer of floor can cover the tops of the nails. You could use little piles of the floor mix in place of the nails. Let these dry some before putting down the floor, so they're hard enough to hold up to the leveling job without squishing.

Drying your earth floor

After you've put down your floor, let it dry completely. This will take a few weeks, depending on your climatic conditions. It's best to stay off it. Protect the floor from direct sun and from freezing. A layer of loose straw will help protect the floor. This will slow down the drying some, so you may want to remove the straw when the conditions don't warrant protection. If you live where it's hot and dry, it might be a good idea to cover the floor with tarps or straw during the day to slow the drying and minimize cracking.

If you have to walk on the floor while you're working on it or before it's completely dry, put down 2 foot squares of plywood and step on them or you'll have footprints.

This is a good time to get firewood, put the gutters up, work on the drains, finish the porches, plant the fruit trees, visit non-cobbing friends and family, or go sit down and read a book.

Sealing an earth floor

The sealant for earth floors needs to soak into the floor rather than forming a hard (crackable) skin on top of the floor.

floor sealant recipe

The basic recipe is boiled linseed oil (if you have an aversion to linseed oil, you could try other oils, possibly coconut?) and solvent (turpentine, mineral spirits or a citrus solvent). The solvent thins the oil and helps it soak into the floor. This sealant looks good on stone and brick too. Experiment!

CAUTION!! Oils and solvents are very flammable and can spontaneously combust. Be careful!

The first coat can be pure oil, because at this point the floor will be very porous. Adding solvent helps it soak in after the previous coat has made the floor less porous.
The second coat can be 3/4 oil to 1/4 solvent. The third coat can be 1/2 oil and 1/2 solvent. The fourth coat can be 1/4 oil and 3/4 solvent. (This last mixture can be used whenever you want to increase the shine of the floor, like maybe once or twice a year.)

putting the sealant on the floor

Let the floor dry thoroughly before sealing.

Warm the mixture in a double boiler **carefully.** Stop before the mixture gets to the point of smoking. Paint it onto the floor while it's still hot. It will absorb better if the floor is warm too.
Put the coating on with a paint roller or a large brush. Use a brush around the edges just like you would if you were painting. Let each application dry, then put on the next one. Apply as much as the floor can absorb and no more. If it puddles on the surface, mop it up.

If you want a really polished look, try heating up beeswax mixed with a little solvent and oil. Paint this on as a final varnish for the floors. Experiment on your floor tests, or in out-of-the-way places on your floor until you come up with a finish you like.

Caring for a cob floor

This kind of floor is softer than what you're probably used to and will require gentleness. Pad the feet of the chairs so they don't gouge the floor. Take your high heels off at the door. Don't drag furniture across it. Use your common sensitivity. One good thing about a softer floor is that it will be gentle on your feet and legs.

Repairing a cob floor

If the floor gets a ding in it, or if one of the cracks starts to enlarge, it's best to patch it as soon as you can. Holes grow when walked on. It's the same philosophy as, "a stitch in time saves nine." Dig the hole out, creating a bigger hole with the edges slanting so that the bottom of the hole is wider than the top.

Moisten the hole well and patch it with the mix that you saved from the original floor for just this purpose. Seal it like you did the original floor once the repair has dried.

If you know you'll be hard on the floor in certain areas, you can put rugs down to protect the heavy wear spots. You may want to make the floor out of something more durable in those places, like brick or stone.

Some other floor surface options

flat stones

These can be set into a silty or sandy soil layer. Put the flattest sides up. The stones can be laid with or without mortar between them. You can use a cob, soil cement, or concrete mortar to fill in the spaces between the stones or bricks. (See the section on mortars in the chapter on foundations, pages 50-51.)

brick

Bricks are expensive if you buy them new. See if you can scrounge up some used ones. Old bricks have become popular and are harder to come by than they used to be. Good luck! Bricks can be used with or without mortar. They can be laid slightly crooked to each other to create a flowing pattern that will suit organic-shaped walls better than the usual parallel rows.

heavy clay

Some earth floors are made thicker and with a heavy clay content. This type of floor is often wetter and is poured into the house. If it is wet enough, it will level the surface itself like water would. Some folks mix the floor right in the house, throwing the ingredients in, adding water, and treading it in place in a big, muddy party.

This kind of floor is allowed to crack up naturally and then the cracks are filled in with a different colored mortar to create a flagstone look. The partially hard floor can be cut into tiles or shapes to give it a controlled place to crack, and then those cracks can be filled. The 'grout' or mortar can be thinned down and drizzled from a pitcher or tin can into the cracks.

This type of floor is usually 4 inches thick, either poured all at once or in two 2 inch layers. It will take a long time to dry - so if you live where the humidity is high, you'll probably want to decide against this kind of floor.

rammed earth

A floor can be made of rammed earth. Cob ingredients (minus the straw) are mixed well and laid down almost dry (ever so slightly moist) onto the floor base. Tamp the earth really well with a flat tamper. A good floor tamper can be made by welding a 12 inch (or less) square flat piece of steel onto a steel pipe, or by screwing a 12 inch (or less) piece of plywood to your wooden tamper. I suggest you make tests in potted plant containers and see what you think. An advantage to this type of floor is that it will dry quickly. Although it is not as durable as a cob floor, it is good under tile or carpets.

soil cement

(See page 52 in foundation chapter for more about soil cement.) Make up test batches as if you are mixing cob, but add different amounts of cement to the dry ingredients. You'll probably want a soil cement floor to be at least 4 inches deep. Try tamping some of the tests. Dry the samples. Did they crack? Are they strong enough to walk on? The surface can be cut, printed, or pressed to look like tiles or shapes. The spaces between the tiles can be filled with a different colored soil cement to look like grout. Soil cement can be used to make tiles. Both these tiles and the poured soil cement can also be used for the floors of patios and porches.

You can use a **rototiller to make soil cement floors** (or even roads) if you have a sandy soil. Figure on making them 5 inches deep. Calculate how much cement to add (6 to 10%) to each square yard or meter.

67

Sprinkle the cement on and mix it all up dry with the rototiller. Then spray it with water and rototill it again. You can tell if you have the right amount of water if you can squeeze it into a ball and it holds together, and if when you break the ball in half, it doesn't crumble. Then rake it smooth and **tamp it well.** Remember to let it cure slowly.

tiles

Tiles make a very beautiful flooring. They are expensive in some places. You can always get seconds or broken tiles and make a mosaic floor. Broken organic shapes look great with the rounded walls. You can make your own tiles out of soil cement, fired clay, concrete or even wood.

wood

To make a loft or second-story floor, simply bury the supporting beams and/or joists into the cob.

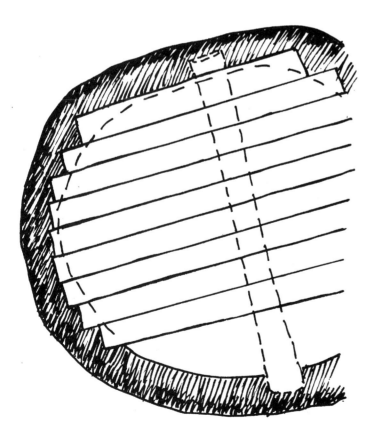

You may want to put the floor in as you build. That way, where the cob wall and the floor meet, the squared ends of the floor boards will be buried in the cob wall. This makes the finishing work easier than cutting curves on all the boards if you put the floor in later. The floor will serve as a scaffolding while you cob the second story. Protect it so it doesn't get wrecked during construction.

I have never built a wooden floor on the ground floor of a cob house. To me that would waste the opportunity to have a thermal mass floor that offers heat storage. It would also use precious wood. If you do decide to make a wood floor you might want to consider the following:

- Floor joists or joist supports can be built into the foundation or walls.

- Wooden floors need lots of ventilation under them. Make the floor high enough so plenty of air can get under there. Make plenty of vents. Cover the vents well so critters can't get in.

- You may want to make the floor high enough to crawl under to fix plumbing or wiring, or to set rat traps.

- The space under the floor would be a good cool place for food or wine storage.

- Remember to add height to the walls to compensate for the amount of height you'll lose by raising the floor.

- I've seen wooden floors laid directly on a concrete slab that has heating coils buried in it. The heat from the pipes keeps the floor dry enough to keep it from rotting. The same thing would probably be true using heating pipes in a cob slab under a wooden floor.

COB GLORIOUS COB!

I suggest you read through this chapter once before you try cobbing.

Photo by Scott Worland

How does cob work?

Soil usually contains sand and clay which are the main ingredients of cob. Cob is sand 'mortared' together with clay and strengthened with straw. Imagine the cob as a miniature mortared, reinforced, stone wall. The sand particles, like the stones in a stone wall, provide strength. The clay serves as the mortar. The straw does the reinforcing job of the rebar (reinforcing steel).

You will be sculpting a semi-liquid that will turn to "stone".

A COB RECIPE

50%-85% sand
50%-15% clay
straw to taste
water

The recipe for cob is pretty flexible. Lots of variations work well.

Materials vary so you'll have to make up some test bricks until you find the best recipe for your ingredients.

The amount of straw is hard to quantify. I like to add as much as I can, until each piece of straw is surrounded by the earth mix. You'll need to stop adding straw before there's too much straw, or the cob will fall apart. The more clay in the mix, the more straw you can add. Try it. You'll get a feel for it. Read on for more details.

Test your soil

Soil is made up of various sized particles: tiny stones, sand, silt, clay, and organic matter. To find soil suitable for building material, first scrape away the top organic layer. Use this top soil for nourishing your garden plants. Examine what's underneath. You will be able to tell a lot by feeling and looking at the soil. Can you recognize sand particles and sticky clay?

Take a cup or two of soil from various potential house sites and from various depths. (Soil samples can vary a lot even a few feet from each other.) Take out any stones or pebbles. Put each sample in a quart jar. Label each jar, recording the depth and where the soil came from. Then fill the jars 3/4 of the way to the top with water. Shake well. I mean really well. Shake that baby! Then let it settle. If your soil has sand, silt and clay in it, you'll get three distinct layers.

The sand is the heaviest and will sink to the bottom immediately as you watch. The silt will settle next, more slowly, and the clay will stay suspended in the water for a while then settle on top of the silt.

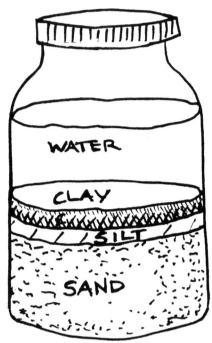

This will give you a rough idea of the ratio of sand, silt, and clay at each spot on the land. If your soil only has one or two components to it, it will be difficult to know what the layer(s) are. See if you can tell by feeling. Make jar tests with other soils until you find one that separates into three layers. Then you can get to know the feel of sand, silt, and clay.

Start out by making test bricks using just the soil from your site. If you need to import materials, stretch the recipe to use as much of the on-site soil as possible. **Too much clay in the cob mix will make it shrink and crack as it dries. Too much sand in the mix makes the cob crumbly and hard to sculpt, and when dry, sand sloughs off if you touch the cob.** If you run into these situations, you'll know you've stretched it too far. Depending on your materials, recipes with as much as 50% clay and 50% sand or as little as 15% clay and 85% sand can work.

Making Test Bricks

To figure out a good mix for your soil, make up some test bricks. Try various recipes and combinations, making sure you label each one carefully so you can reproduce the best ones.

Make your first couple of test bricks out of the soil from the site or from where you've dug out for the foundation. If you're lucky, you may have a soil that works for cob as it is without adjusting the sand/clay ratio. After the jar test, you'll have some idea what you might want to add for the other test bricks. Make tests adding more and more sand until you've stretched it to the point of **too much sand. You'll be able to recognize that point because the mix won't hold together well and will be hard to form into bricks.** The sand will slough off when you rub the dried brick. You can take the experiment to the opposite extreme and add too much clay so the brick cracks while drying. You will learn a lot doing these tests. Make three samples of each sand/clay mix, adding different amounts of straw until you get a feel for how much you need.

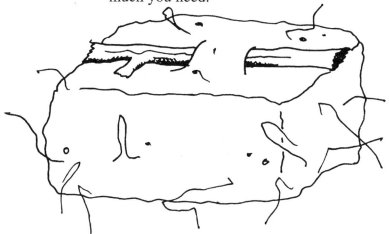

One easy way to check the clay shrinkage in your test bricks is to bury in a stick the same length as the brick. The cob might shrink but the stick won't. You can write the recipe for that brick on the stick with a permanent felt pen.

When the bricks have dried completely, have a look at them. Eliminate the ones with the biggest cracks because they have too much clay in them. Does the sand come off when you rub the bricks? If so, those have too much sand. Try breaking the others. You can drop them from one or two feet up to test their strength. The ones that are hardest to break are the best recipes. Leave them out in the weather or simulate some rain and observe which ones are the most resilient. Analyze what amounts of straw made the strongest bricks. Now you have a good basic recipe. Lots of variations will work just fine, so **don't be too much of a perfectionist about all this.** You may want to stretch the recipe so you can use as big a proportion as possible of the soil you have on-site.

If you want to reassure yourself and do a more extensive test, build a garden wall, an outdoor sculpture, or a bench, and observe how hard it dries and how well it weathers. Remember, your walls will be protected from the rain by a roof.

More details about cob ingredients

SAND

Sand particles are the main building blocks of cob. Ideally, each particle will have **rough angled sides** so the sand pieces interlock with each other, helping to hold the cob together. If the sand particles are rounded and smooth like ocean beach sand, they'll be more likely to slide off each other.

Sand that's made up of **various sized particles** makes a strong cob mix. The pieces can vary a lot. I think some pebbles and small rocks also strengthen the cob. Don't be too fussy about this. If in doubt, make test bricks to see if what you have will work.

If the soil from the site is short on sand, have a look around your property. Soil can vary a lot even within just a few feet. Often the movement of water in streams or rivers, past or present, has left sand deposits somewhere.

If you need to buy sand, go to a concrete supply place and ask for **coarse concrete sand or 'reject' sand.** In most places it's pretty cheap. For an extra fee they will usually deliver it to your site in a big dump truck, if you have a decent access road. Or you can haul it yourself, a truckload at a time. The gravel yard will load it for you. Don't let them overload your truck! If you can't drive all the way to the cob site, take it as far as you can, then load it directly into your wheelbarrow or garden cart to carry it the rest of the way.

It can be tricky to figure out **how much sand** you'll need. It depends on the size of your house, the thickness of the walls, the number and size of the window and door openings, and how much sand you'll have to add to your soil to get a good mix.

Remember: if you plan to make an earthen floor, sand will also be a major ingredient of your floor. You can roughly estimate all this by volume. If you're getting the sand yourself as you need it, it will be less important to know the total amount. If you're having sand delivered, ask what the minimum delivery is. If that seems reasonable for your plans, order that amount. You can always get more later if you need it. If there's extra, you can probably use it for landscaping, gardening or future building projects.

CLAY

Clay's job is to hold things together. Clay is one of Mother Earth's magic mysteries. Legends of many cultures tell how the first people were made out of clay.

Clay is made of very small platelet-shaped particles held together by the friction of their surfaces. The particles are small enough to remain in suspension in water for a little while. **Clay expands when wet and shrinks when dry.** Some clays change size more dramatically when wet or dry than others do. Clays also differ in stickiness and ability to glue things together. Most soils that I've come across, in eastern Australia, northern New Zealand, and the northwestern United States, have had plenty of clay in them so I haven't had to import clay.

If you have very sandy soil and need to add clay, you'll have to find some. Gathering and adding clay can be a lot of work so if you have to do this, figure out a cob recipe with as little clay as you can get away with. You don't need much, especially if you can find a good sticky clay and add a minimum of straw.

Where to look for clay

Often the deeper you dig down, the more clay you'll find in your soil. The sides of streams or rivers are likely places to find clay deposits because the water has worn into the lower layers of soil. Look along the sides of roads where the earth has been cut. Hopefully you can find some in a location where you can load it into your truck.

Clay is used by tile and brick manufacturers and can be purchased cheaply from them.

If you need to add clay to your soil, either mix the clay in water first, or dry it and crush it to a powder before adding it to your cob mix.

Adding the clay wet

Soak clay-rich soil overnight (at least). This will help it to mix well with the water and make it easier to break up the clay. If your soil has almost enough clay, you can separate clay from some of the soil that you won't be using for the cob, and add it to your mixes to raise the clay content. Use the same principle as the jar test: put soil into a big garbage can, barrel or pit, fill it with water, and stir it a lot with a hoe or by mushing it up with your feet.

Straw bale mixing pit

One easy way to make a pit is to build the sides of the pit with straw bales and line it with a tarp. The clay will become suspended in the water and then settle out on top of the dirt. You can scoop off this purer clay and add it to your mix. Straining helps break up those stubborn lumps of clay. When you add wet clay, you can thin it down with water until you can use the clay "soup" for the liquid _and_ clay proportions in your recipe. You may need to add a little more water directly to your mix if it is too dry.

Powdering dry clay

If you add dry clay to the sand, you'll need to powder it first and mix it with the sand while dry, then add the water. This is a dusty job and bad for your lungs, so wear a mask. The dry clods of clay can be crushed with a tamper or stomped on with boots. It can be grated through a wire-mesh screen. Driving your car over the clay clods with the windows rolled up is a good way to avoid the dust. If you plan way ahead, the clay can be left for a year or two exposed to the weather so that the sun, rain, and freezing will help it to break down naturally.

Silt

Some silt in your cob mix is OK. Cob made with a large proportion of silt is weaker but if that's all you have, try it. Make a test wall first and see how it holds up. Your test wall could be a garden wall, a bench, or any other small structure you can think of.

Silt has larger, more gritty particles than clay does. One way to help you establish whether the material is clay or silt is to make it into a dough ball and cut it with a knife. If it's clay, the cut surface will be shiny - if it's dull, it is silt. Organic matter thrives in silt. If there's lush plant growth on the soil, it probably has a lot of silt in it and would be wasted in cob. It's much better used in the garden. You may find a layer of clay under the silt.

STRAW

Straw adds **tensile strength** to the cob.

If you have a clay-rich soil and limited sand, you can use a heavier clay proportion by adding extra straw. If you have a very sandy mix, it won't be able to hold as much straw. That's OK.

Straw is the stalks of grasses or grains left after the nutritious seed heads and leaves have been removed. It's cheap stuff, often baled for easy handling and transportation. The fresher it is, the stronger it is, so avoid old straw that's started to decompose. Make sure to keep the straw dry until you're ready to use it. **Do not use hay.** It decomposes.

Some people think certain lengths are better than others but I just throw it in as it comes out of the bale and it seems to work great. The straw will soak up some of the water in your cob mix, drying it slightly.

Get at <u>least 10</u> bales for making a small cottage or for each large room. More is good.

Extra straw bales are wonderful to have at the building site. They serve as scaffolding supports, ladders, walls for clay-making pits, chairs, backrests, tables, even beds. Eventually they'll become great garden mulch or erosion control.

If you're an ambitious purist, you can gather your own grass stalks or experiment with other plant fibers.

Water

I usually use plain fresh water in the cob mixes. I have tried all of the following ideas and they all seem to work fine.

- Some folks say rainwater is best.

- You can soak manure in the water and use that water for making cob.

- Some olden-day recipes say that to strengthen the cob you should add a little bit of lime putty to the water (3%). (See page 162 for how to make lime putty.)

Tarps

When I first started making cob, I used to turn it with a garden fork. **Mixing the cob on a tarp is a much easier way to stir the ingredients and save your back,** so get a good supply of tarps. Any big tarp of sturdy material will do (7x9 foot or bigger). Those awful woven blue plastic ones work well. They're light and the dirt doesn't stick to them much, but they do disintegrate after a while. The fibers from the shredded tarps can be added to the cob mix for reinforcing and to keep them out of the landfills, tips or dumps. Old awnings from recreational vehicles are good for the job too, and last longer than the tarps. RV repair shops are often happy to get rid them.

Tools you'll need for cobbing

- cobbing tools: sticks and stones that fit your hand
- shovel and spade
- pick
- tarps
- machete
- squirt bottle (put a hole through the lid with a nail if you don't have a squirt nozzle lid)

- hose with nozzle
- wheelbarrow
- 3 foot level, wedge and tape to attach the wedge to the level (See page 87.)
- buckets
- plumb bob or something heavy tied to a string to help you see a vertical line
- whisk broom
- ladder

Here's how to make cob

Cob is mixed with a combination of stirring and compression. In my experience, the best way to achieve this combination is with feet and tarps.

Clear a few places at the site for mixing areas. The closer they are to the foundation, the less energy it'll take to get the cob to its destination. Inside the building is a good place for at least one mixing area. Leave enough room to walk around the mixing sites easily. Use your brain to minimize work. If you've imported ingredients, pile them in sensible places to streamline the process. For example, if the sand pile is two steps closer to the building (but not crowding walkways) you'll save miles of steps during the building process.

If you need to add sand or clay to your soil, you can measure out proportions by the shovel full, or by the bucket full until you can do it by eye and by feel.

Throw some soil on a tarp

Break up the clods with your tamper or your shoes.

Adding sand
If you are adding sand, shovel the sand onto the tarp and stir it into the crushed soil until the two are well-mixed. You can do this by pulling up the edges of the tarp and rolling the ingredients around on the tarp.

Adding clay
If the soil you are working with is so sandy that you have to add clay, you can either add clay soup or add dry powdered clay to dry sand. Wear a mask when adding and stirring dry clay to your mix.

Stirring and treading the mix

Stirring and treading the ingredients well is a big part of the magic that turns your ingredients into cob.

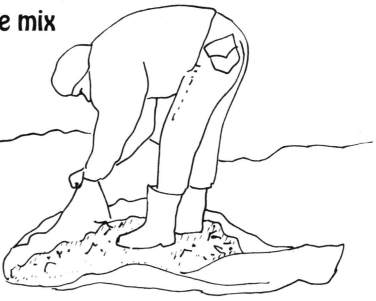

To stir, **stand on one edge of the tarp and lift up the tarp on the opposite side of the mix. Pull it _towards_ you, turning the mix over onto itself.** Don't lift the edge of the tarp at your feet or push the mix to turn it. That's too hard on your back.

Then **add water.** You're after a cookie dough consistency. **Tread on the mix with your feet.** It's easier to feel the mix with bare feet, and for some mysterious reason, bare feet mix the cob faster than shoed feet. Don't despair, your feet will return to nature and toughen up quickly.

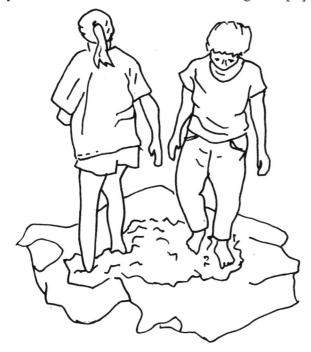

Stir the mix often by pulling up the edges of the tarp, while you're treading. It's impossible to tread the mix too much, so go for it. You'll notice how it changes plasticity as you tread on it.

If your soil has a lot of **sharp rocks** in it, you can gradually toughen up your feet or give in and wear shoes. Smooth soled ones work best because the cob doesn't stick to them so much. Another option is to sift the dirt through a 1/2 inch wire screen. This is a lot of work and puts dust into your lungs. Wear a mask if you do this! Breathing dust is bad for you! **Be aware that the clods that won't go through the sieve easily are the hardest and best pieces of high clay-content soil.** These can be forced through the sieve, or crushed and then sieved. Or these precious little lumps of clay can be soaked overnight, then put through the sieve wet or mushed up by foot. You can save the gravel that you end up with after sieving for your drains, or put it back into the cob after you've finished mixing it.

When your mix begins holding together in a giant loaf when you turn it with the tarp, it's time to add straw.

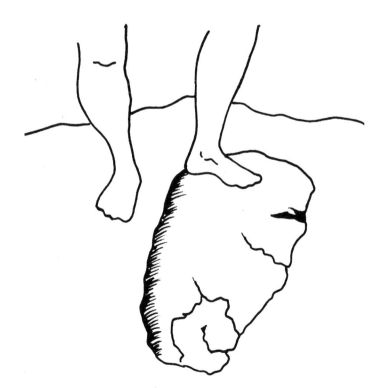

Adding straw

When you add straw to the mix, it will dry it out quite a bit. If the mix is too dry to add lots of straw, you can either add less straw or add water. You will soon get a feel for the consistency you're after. Remember this ancient knowledge is imprinted in your cells. Trust yourself. Experiment.

Flatten the cob loaf down by treading on it. Grab an armful of straw and sprinkle some onto the mix. Step on it so it sticks into the cob a little and pull the tarp up, making a cinnamon roll of dirt-dough and straw-sugar. Tread on it until it's flattened out into a pancake again, and add more straw. Make another cinnamon roll. Do this three or four times until you have enough straw in the mix. Too much straw will cause the mix to fall apart. Try adding a lot to a small mix so you can see what too much straw feels like.

Cob too dry?

A cob mix that's too dry will be crumbly, won't want to stick together very well, and won't take straw well. Add water and stir until the mix is the consistency of cookie dough. Too much sand or too much straw can also cause a crumbly mix.

Cob too wet?

When you build with a mix that is too wet, you'll see that it won't keep its shape because it can't hold up its own weight. It will start to bulge out on the sides of the wall. The technical term for this is "ooging". (All cob oogs. The wetter the cob, the sooner it oogs.) If it's too wet, add a little dry sand, dirt, or straw and tread it in. If you add a large amount of dry stuff, use roughly the same ratio of dirt and sand as your original recipe. You can also dry a mix by treading it into a pancake so the air can get to it, and letting it sit for a few hours or overnight. You'll quickly get a feel for it.

Letting the mix cure

Because **it is so much easier to mix the cob if it's on the wet side, you may want to do it that way and let the mix sit for a day or two** to dry out before you put it on the walls. When mixes sit, something magic happens and they become more elastic and easier to work. You'll get a feel for how wet to make them and for how much they'll dry during a night. If they're too wet, uncover them and let them sit a little longer. Once you get into the swing of cobbing, you can make mixes on all your tarps at the end of the day when you're warmed up. In the mornings you can get inspired and warmed up by cobbing onto the walls.

Curing is not necessary. It's fine to make a mix and put it directly onto the wall. I suggest you try letting a batch of cob sit, though, to see if you can feel an improvement.

Avoid leaving the mix covered for too long because the straw will start to rot and stink, and lose some of its strength. If you do this accidentally, it's no big deal, put it on the wall anyway. It will stop stinking as soon as it dries.

Other ways to mix cob

In the old days in Europe, **farm animals** were often used to mix the cob. The ingredients were shoveled into a circle around a post. The horses or oxen were tied to the post and walked round and round treading the cob mixture.

In recent times people have used various machines to make big batches of cob, and to speed up the process. **Caution:** if you let a batch of cob sit wet for too long, the straw will start to rot. It's a good idea to only mix big batches when you know you'll have time to get the cob onto the wall fairly soon. One strategy is to mix a big batch of the clayey soil and sand and add the straw to only the portion of the mixture that you know you can use within a week or so.

Cement mortar mixers will mix cob if you can handle the fumes and noise. As with manual mixing, the clayey soil and sand is mixed first, then the straw added. The cob has to be mixed wetter than with other types of mixing to allow the cement mixer to turn without too much stress. This means the mix may have to sit a day or two to dry to the right consistency. Straw can be added manually to the mix.

Tractors or four wheel drive vehicles can make quick work of the cob mixing job. I've seen the process happen unintentionally in peoples muddy driveways. Simply pile the soil and sand and drive back and forth over them. Add the straw after the other ingredients have been mixed well, then drive over the mixture again until it's well stirred.

Heavy earth moving machinery can be used to mix up big batches of cob. Front end loaders are great! They can do the job of compiling and mixing the ingredients. The finished cob mix can be scooped up in the loader and raised to the height of the wall. You can climb into the loader and hand-sculpt the walls without having to lift the cob up onto the wall by hand.

Putting the cob on the wall

Drag the mix on the tarp to the spot where you want to build. Anywhere is fine. If the foundation is low enough, two people can roll the mix off the tarp and up onto the foundation. If the foundation or walls are too high to roll the cob up onto, you can make a ramp to drag it up or lift it by handfuls, tubfuls or buckets onto the foundation.

If you haven't already put the pipes in for the electricity and plumbing, now is the time to do it!

Cobbing by foot

sometimes called 'pisé', (pronounced 'pee say')

This is a method of squishing the cob mix down onto your foundation by stepping on it. This pushes the cob into the spaces between the foundation stones.

Keep adding cob and compressing it together with your weight.
When the sides start bulging out, get a friend to use their feet
for a form.

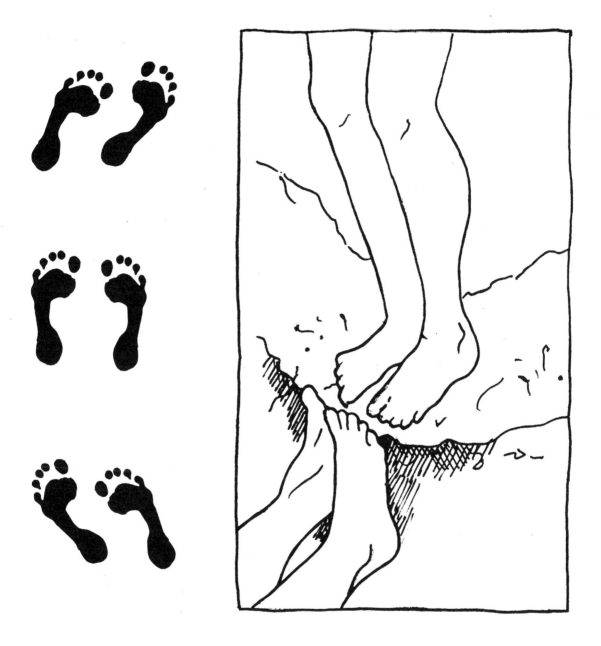

Get your weight off the cob when it starts **bulging out too much on
the sides, technically called 'ooging'**. Let the cob harden for a day
or two, until it can support your weight. You can cut off the oog later
with a big knife or spade, before it's hardened completely. If you
like foot cobbing and have good balance, you can do it all the way to
the top of the walls. Make the walls thick enough so you can come
back and tidy them up by cutting off the rough, protruding cob with
a machete or spade.

Cobbing by hand

Cobbing by hand is more gentle and controlled, creating more finished walls. Hand cobbing will allow you to build up a little further before the cob starts ooging because your weight is not added to the wall.

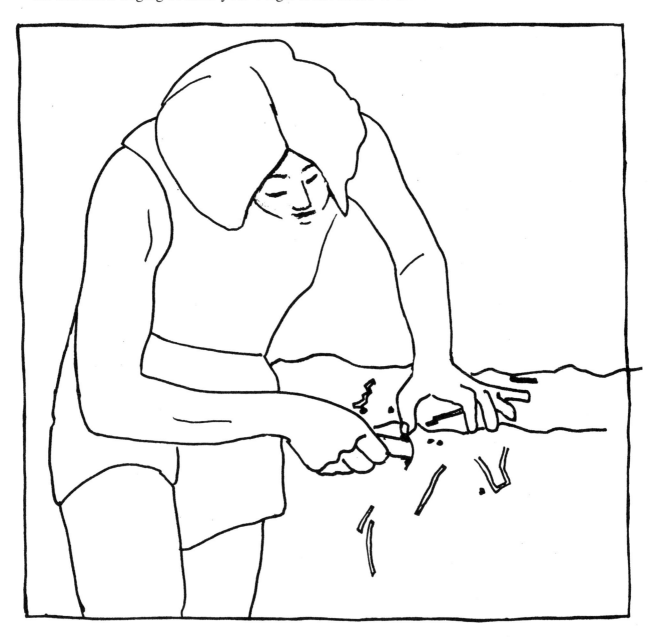

To loosen the cob from the tarp and make it easy to pick up, roll the mix in the tarp like you would for stirring. Lift handfuls of cob onto the walls and massage it all together with your hands and a stick.

Thumbs are a great tool for the job but they do get tired, and thumb nails get sanded-down from shoving them into the cob. Give your thumbs a rest by replacing them with a stone or stick part of the time. You can use your palms and knuckles too. One hand can be the form as the other compresses the cob against the form and incorporates it into the cob below it.

It's easiest if you can keep your upper body weight over your arms, using gravity to help you push and massage the cob together.

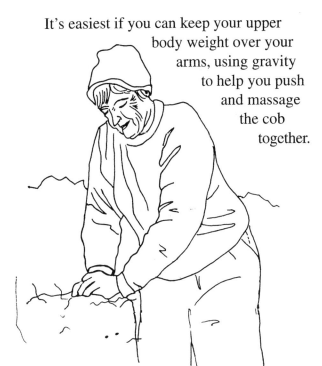

It's important to make your house one massive hunk of cob, avoiding weak connections between 'layers'. When adding cob to a wall, you may need to re-wet the last application of cob so it's soft enough to massage the new cob into it.

Do not smooth out the surface of the cob too much or it will dry as a hard skin, which will slow the drying in the wall beneath it.

When you're cobbing, you are creating a surface for future cob to be built onto. Make sure you get into the habit of making a flat surface on the top edges of the wall, not a rounded one, so the next "layer" of cob has ample surface area to sit on comfortably, instead of sliding off.

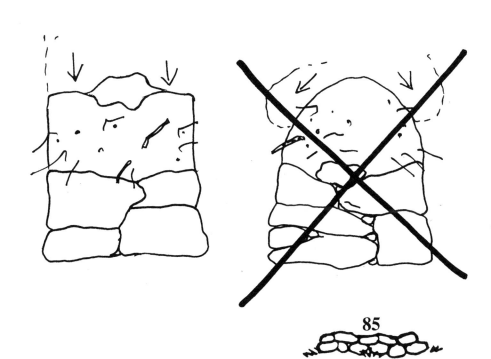

Tapering your walls and how wide to make the top of the foundation

Make the **bottom of the wall wider than the top.** I usually make the interior side of the wall pretty straight up and down. It's a matter of taste. If you want your walls to grow organically, use your common sense. You can angle the wall for the first foot or so and then make it straight up, imitating a tree trunk or the interior of a cave.

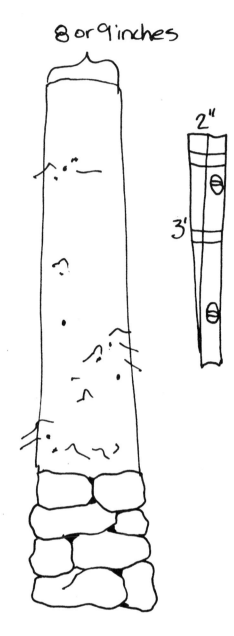

8 or 9 inches

2"

3'

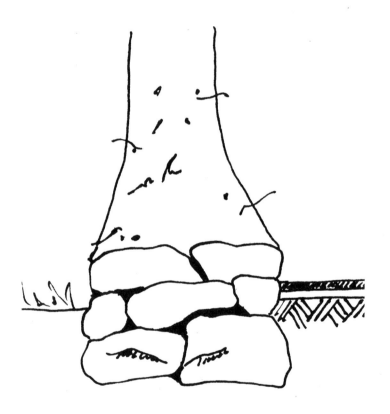

The wall has less and less weight to support as it goes up, so it doesn't need to be as thick as it gets nearer to the top. You can **taper the outside of the wall**. Tapering cuts down on the amount of cob you'll need and makes the house that much lighter. The angled shape is a stable shape. **Taper two inches for every three feet of height.**

On average, a wall will need to be 9 or 10 inches wide at the top, tapering out 2 inches for every 3 feet of height. If your wall is 9 feet tall, start by taking the 9 inches you'll want at the top and add 6 inches (2 for every 3 feet of the height) which comes to 15 inches. Add another inch or two for good luck. So that's 16 inches at the bottom of the cob wall and at the top of the stone foundation. If you're planning two stories, use the same formula. **These calculations are vital for deciding the width of the foundation.**

Cut a wedge measuring 2 inches x 3 feet out of styrofoam or wood and tape it to one side of the level. Leaning the wedge side of the level against the outside wall, get the bubble to read level. The edge of the wedge will then show you the taper you're after. Cut off the bumps on the wall to form the plane you want. Use the other side of the level for your interior vertical line.

Keep an eye on the outside planes of the wall as you cob.

As you build, you will be **creating the vertical planes on the sides of the walls** that you will always see. As you add cob, keep an eye on both the inside and outside plane of the wall. Cob has a mind of its own. It's surprising how hard it is to keep the walls growing the way you want them to by just using your eye. If you don't check a level often, it's easy to build a lot of unnecessary width into a wall before you realize you're going wonky.

Cobbers tend to push the wall away from themselves as they massage new cob onto the wall. This sometimes distorts it towards the other side of the wall from where they're working. It's helpful to cob with a friend, with one person inside the house and the other outside, working opposite each other, keeping an eye on the vertical plane on 'your' side of the wall.

When working on a tightly curved section of the wall, work mostly from the **outside** of the curve. Cob curves tend to lean outwards as the wall goes up.

Ooging

If you keep adding a lot of cob at once, it will eventually start to oog or bulge out on the sides, pushed out by its own weight. When that happens, stop cobbing in that spot and move on to a different section of the wall. If you have a big house-raising party and you manage to get to the point where the whole wall is ooging, pat yourselves on the back, make mixes in all the tarps to cure for tomorrow, and call it a day. Unless you live in a very humid climate, the wall will be dry enough the next day to chop off the oog and keep cobbing.

Over build the sides and trim them later

It's much easier and stronger to chop off cob than to add it to a dry, or even a partly dry, surface. It's better to make your walls too thick and bulgy than too thin and dippy.

A day or two after cobbing, the walls can be checked with the level and trimmed to give you a more exact shape. Do this in the mornings before you start cobbing. A machete, meat cleaver, or spade are good tools for shaping the walls. Chopping cob will dull your tools, but they seem to work just fine without sharpening. A 2x4 held on its side and scraped across the wall helps to make a flat surface. When you're doing the trimming, stand back occasionally and look at the walls from a distance to give yourself perspective.

The cob that you chopped or scraped off can be re-wet, retread a little, and recycled back into the wall, or thrown into an almost-finished new mix and tread in a little.

Another totally different approach is to build the whole house, leaving the sides of the walls all lumpy and messy. Then after the walls are built, wet them down really well and carve them into their final shape all at one time. A hoe or shovel are good carving tools to use along with your big knives.

Putting the cob to bed at night

At the end of a cobbing day, remember to make sure you haven't left any dips in the sides of the walls and poke holes in the tops of the walls to allow for drying and re-wetting. The holes serve another purpose too. They provide a key for the new cob to hold onto. When you continue cobbing, push the fresh cob down into the holes. When you've finished cobbing for the day, poke finger-sized holes with a stick in the top of the wall, especially along the edges. Make them 3 or 4 inches apart.

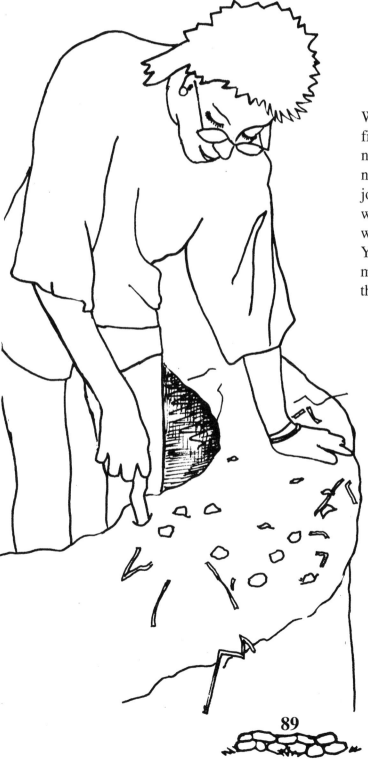

When you want to cob again, you can fill the holes with water if the old cob needs re-wetting. A hose with a spray nozzle is perfect for the re-wetting job. Keep some squirt bottles full of water around the cob site for re-wetting smaller areas as you cob. You will soon get the feel for how much to wet it so that you can work the new mix into the old.

Control the wall drying

It is easier to get the previously applied cob and the new cob to stick together if the moisture levels are similar. This means that you are riding the fine line between **keeping the top of the wall as moist as possible so the next cob layer will adhere to it, and letting the lower part of the wall dry enough to support the weight of more cob.**

Cover the tops of the walls with dampened straw, burlap sacks, and/or tarps and kiss it good night.

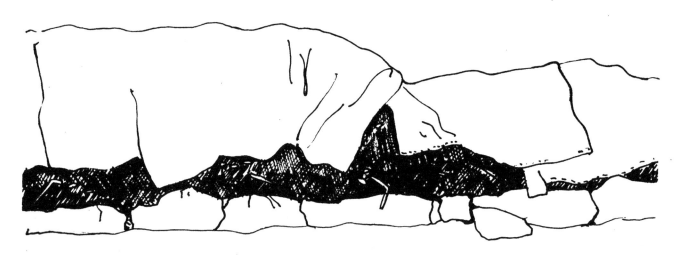

If you suspect heavy rain is coming, you can cover the walls with tarps or sacks. To speed up drying, you can place lots of dry straw on top of the wall under the tarps or sacks. If it's likely to be windy, weigh down the covers so they won't blow off. The rain won't hurt your wall seriously but it will, obviously, slow down the drying time.

It's a good idea to keep your fresh walls out of strong direct sun and harsh winds, which will dry the outside surface too quickly, and can cause cracks on the dried surface as the inner cob dries and shrinks. Protect the walls with shade or by covering them. If you're cobbing where the temperature drops below freezing at night, pile straw bales or loose straw on the sides and on top of the walls to prevent them from cracking up. You could use old sleeping bags or anything that insulates. Take the covers off during the day so the walls can dry.

If you'll be back to do some more cobbing in the next few days, stop working on the walls while you still have enough energy to make up some cob mixes to cure. It's a delight to have mixes ready at the start of a new cobbing day. Because these mixes will be curing and drying out for awhile, you can mix them extra wet which is easier and quicker.

I also like to tidy up so when I return to the site it feels welcoming, and I can find my clean tools.

It's as simple as that! Keep cobbing and sculpt your home!

What if the cob dries out completely before you get back to it?

It's OK. Fill the holes in the top of the wall again and again with water until the cob is rehydrated. Be careful to really incorporate the first new layer of cob. Use a stick and push the new cob hard into the old. Another trick is to sew the layers together with straw, pushing pieces of straw through the new mix into the holes in the old cob.

If you know that you'll be leaving the partially finished building for a long time, there are a few things you can do to make it easier to connect the next 'layer' of cob.

• Leave the top of the walls rough, with lots of humps and holes and valleys.

• Make the re-wetting holes extra wide and deep.

• Cover the top of the walls with wet burlap sacks and let the rain fall onto the walls.

• You can push small sticks or bamboo into the cob on the tops of the walls 4 or 5 inches down, leaving 4 or 5 inches sticking up. Put them every foot or so and poking out at different angles. The sticks will be buried in the new cob when you continue building.

Push straw into the tops of the wall leaving some of the straw sticking out to help knit the next cob application to the previous one.

Patching an already dried wall

Plaster adheres beautifully to cob and will cover minor dips and smooth out the overall contours of the walls. But if you want to make a more major change and add new cob to the old, start by wetting the dry cob as much as possible. Drill or carve holes in the hardened cob for the new cob to hold onto. You can hammer some old nails partway into the dry cob to help support the cob you're adding.

How to destroy dry cob

If you want to add a window or door, or change something in an already dry wall, get out the battering ram. You will soon have a new respect for the strength of cob and a healthy appreciation for taking the time to plan carefully. A 3/4 inch steel pipe can be sledge hammered through the wall many times where you want an opening. Wetting the wall again and again helps. Try an axe or hatchet to chop out what's left. It is much easier to fill in an unwanted door or window than to chop one out!

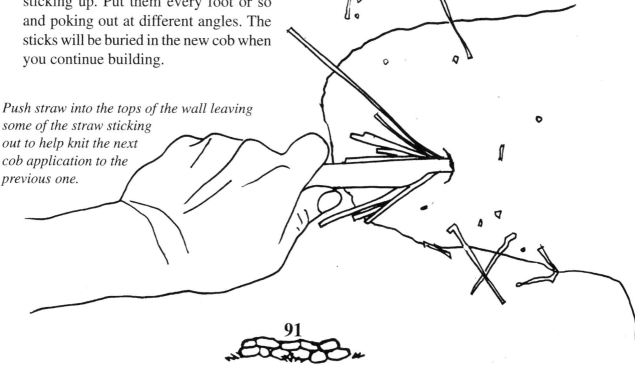

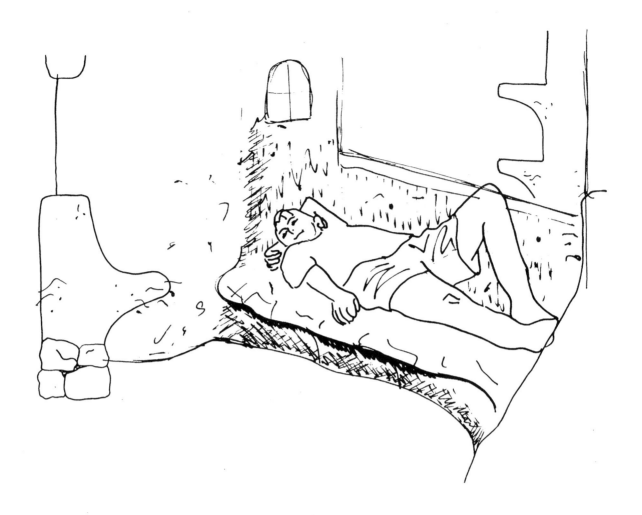

Sculpting cob shelves and furniture

Cob is amazing stuff. It's so strong you can sculpt it out from the wall to create shelves and benches that project into the room (this is called a cantilever). Built-in surfaces for storage and seating maximize your interior space. Make yourself a beautiful window seat by cantilevering a bench out at the bottom of a big window. When you build indoor furniture consider the final floor height, so the seats are at a good position. You can copy the measurements and angles of one of your favorite chairs or couches to help you create a comfy seat. If you want a wide seat or shelves, you may want to build up the thickness of the wall or foundation under the cantilever before you start sculpting it. This will save you time because making a cantilever is time consuming, careful work.

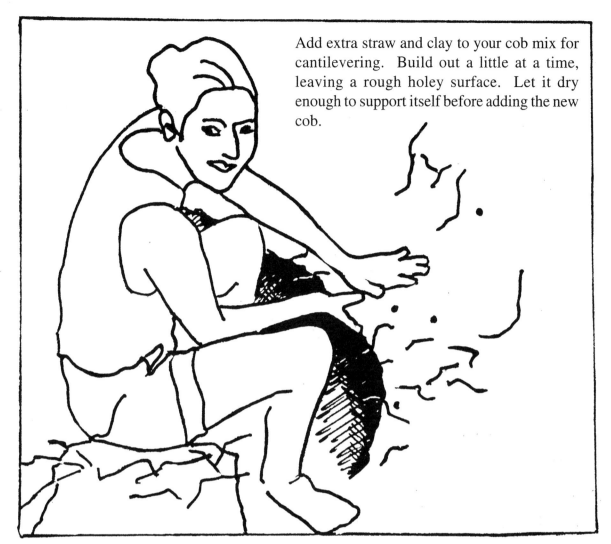

Add extra straw and clay to your cob mix for cantilevering. Build out a little at a time, leaving a rough holey surface. Let it dry enough to support itself before adding the new cob.

Take extra care to make sure the new addition is well attached to the last cob. Cantilevering takes patience and a little practice. Sometimes the cantilever falls off. Don't panic, keep at it, just add on a little slower. You can stick little sticks or straw into the cantilever, sticking out into where the next cob will be added. This will help hold the weight until the cob hardens and add to the tensile strength.

If you use something to temporarily hold up the cantilever while it hardens, like a bucket or straw bale, make sure you remove it within a day, or it'll get stuck as the cob dries and shrinks.

You can always chop away furniture and shelves if you decide you don't like them later on. It's possible to add cob furniture after the building is made, but it will be easier and stronger to do it as the building grows.

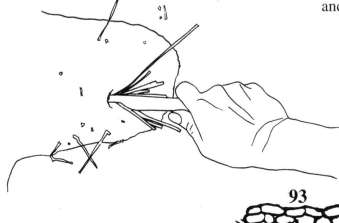

93

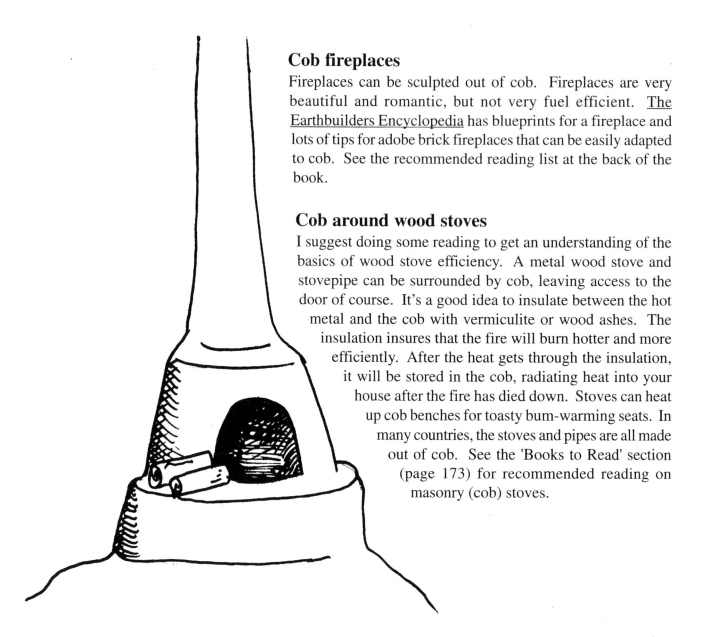

Cob fireplaces

Fireplaces can be sculpted out of cob. Fireplaces are very beautiful and romantic, but not very fuel efficient. The Earthbuilders Encyclopedia has blueprints for a fireplace and lots of tips for adobe brick fireplaces that can be easily adapted to cob. See the recommended reading list at the back of the book.

Cob around wood stoves

I suggest doing some reading to get an understanding of the basics of wood stove efficiency. A metal wood stove and stovepipe can be surrounded by cob, leaving access to the door of course. It's a good idea to insulate between the hot metal and the cob with vermiculite or wood ashes. The insulation insures that the fire will burn hotter and more efficiently. After the heat gets through the insulation, it will be stored in the cob, radiating heat into your house after the fire has died down. Stoves can heat up cob benches for toasty bum-warming seats. In many countries, the stoves and pipes are all made out of cob. See the 'Books to Read' section (page 173) for recommended reading on masonry (cob) stoves.

Relief sculptures on the walls.

Cob is a medium that invites sculpture! It's super flexible. You can make relief sculpture as you cob. You can also make artistic dips and holes in the wall. Or you can cob a bulge onto the side of the wall and when it's hardened a little, carve it to the shape you want. You can add a sculpture to a dry wall if you rough it up well, re-wet it, and add some nails pounded part way in to help the fresh cob stick. When adding large sculptures to an already dry wall, use large nails. You can add one part mushed up newspaper pulp to one part cob to lighten the weight.

If you decide you can't live with your sculpture or you want to change it, go for it. One of the wonderful things about cob is you can chop and change it indefinitely.

Burying in shelves, counters, and loft floors

Any wooden shelves, shelf supports, counter tops, or wood to support cupboards or lofts can be embedded into the cob as you build. It's much easier to sculpt cob around wood than to cut the wood later to the organic shape of the dry cob. Everything that you build in as you go will save you a lot of fussy wood cuts later. You may want to keep the wood clean by covering it with cloth or plastic while you build. The counter, loft, and sturdy shelves can serve as scaffolding. Make sure the cob has hardened enough to support the protruding wood with your weight on it before you stand on it.

If you choose to put your counter or loft in later, you can leave a little cob shelf (2 or 3 inches) to set it on.

To support something small like a counter or shelf, compensate by making the walls a little thicker below the ledge. For a loft, you might want to make the whole wall a little thicker from the foundation up. You can cut the counter (or floorboards) to fit and set it on the ledge.(See page 68) Fill any gaps between the counter and the wall with cob or plaster.

Counters may need extra 'legs' to support their outside edge. These can be made out of wood or you can build cob pillars. These can also support shelves under the counter. If you want to attach cupboard doors, bury in a piece of wood to attach the hinges to. Remember to add something to help key

For two-story houses, you can save space by burying pieces of wood for steps that protrude out from the walls into the room. The cob is very strong and can easily hold the wood in place. This stair system has been used for centuries but is not very safe for little kids and will not pass any building code.

Making niches or cubbies in the walls

Niches are recessed spaces built into walls to create an altar, bookcase or a place for candles. These are fun to sculpt in as you build. If you finish a wall and then wish you had built a niche, you can carve one out later. This is much easier to do if the cob is still wet.

If you bury colored bottles though the back wall of the candle or light fixture niches, you'll see the colored light shining through the walls when you're outdoors.

The tops of the niches can be treated like any other opening. They will need a lintel or an arch. If you make niches less than a foot wide, you can forget about the lintel or arch rule and let your imagination go.

The thinner walls in the back of the niche will be less insulating than the thicker parts of the wall. If you live where it's really cold, you may decide to forget the niche idea. Books are good insulation so you could make a bookcase niche and keep the shelves close together and full.

SCAFFOLDS

You'll need something to stand on as your walls grow to keep
your upper body weight over the wall. Standing on straw bales
or on planks that are resting on straw bales works well for the
lower part of the wall. Only stack the bales two high; three is
too precarious. The straw bales against the walls will slow down
the drying so you may want to move them once in awhile to
give each part of the wall a chance to dry out.

It's OK to stand on the partially dried window sills, cantilevers, and built-in furniture. Use your judgment. You can straddle the wall, sitting on it like a horse.

As your walls get higher, pieces of wood can be buried through the walls to support the scaffolding planks, both inside and out. Put the supports into the walls at least a foot from an opening.

Four or five feet off the ground is a good place for the first scaffolding supports. Depending on how tall your walls are and how agile you are, you may want to put more supports every two or three feet. Be safe, rest strong planks on the buried wooden supports and attach them well. Add some additional sticks to help hold up the outsides of the scaffolding supports. (See illustration.) You can use a ladder or straw bale steps to get up onto your scaffolding. The cob can be piled into dishpans or tubs and set up onto the scaffold, then lifted up onto the top of the wall.

When you're finished cobbing, the supports can be left in the walls to permanently carry the weight of shelving or cabinets. In many countries, the scaffolding sticks are left in the walls as decorative ladders and to stand on during replastering later.

If you want to remove them, they can be sawed off and plastered over or they can be knocked out of the walls with a sledge hammer, and you can fill the hole with cob before you plaster.

On the sunny, glass-intensive wall, it's likely that you won't have enough cob between the windows to hold embedded scaffold supports. A simple scaffold can be set up using two stepladders with the scaffolding planks running between the ladders, and resting on the ladder rungs (not on the top of the ladders!) Make sure the ladders are well-balanced where they stand.

Electric wiring

If you haven't put the access pipes for the electric wire and plumbing through the foundation, make sure you set them on top of the foundation and cob them in place. If you haven't decided exactly where the access pipes should go, just put in extra ones and fill the ones you don't need later with cob. (See more about this in the chapter on foundations, page 43.)

Because cob is completely fire proof, electric wires can be buried right into the cob walls or put in conduit pipes and buried in as you build. Bury them at least an inch deep in the wall to avoid chopping into the wires later when you do the final shaping of the walls. Mark where the wires are buried in the cob so you can find them in the future.

To hold the electric receptacle boxes in the cob, get a piece of wood a little bigger than the box, and nail or screw it to the back of the box. Cob it in where you want it. If you'd rather deal with finishing the electricity later, leave a hole and, when you get to it, embed the box into the hole with cob.

Another approach to installing the electric wires is to carve a groove out of the cob where you want the wires to be, either when it's wet or dry. Lay the wiring in the grooves after you've finished the walls and just fill the groove with cob or plaster afterwards.

Other things to think about
Termites and silverfish

I know very little about termites, but would like to pass on an idea I heard in Australia. The concern was that the termites might bore up through the cob and get at the wood in the frames and the roof. One suggestion was to place termite mesh through the wall with tin strips welded or soldered to the sides, protruding out of the wall about 3/4 of an inch both inside and out.

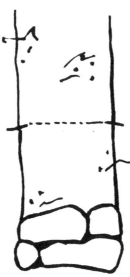

Termite mesh is a heavy duty mesh available where termites are a problem. It has holes too small for the termites to get through but will allow the walls to breathe. The solid strip of tin makes it easy for the home-owner to see whether the termites have built their own little cob tunnel over the tin. The tunnels can simply be knocked down to discourage the termites from getting up the wall. Any other ideas?

In Australia, I talked with an old friend who lived in a mud brick (adobe) home that she had built herself and had been living in for 12 years. She said she had a problem with silverfish (little destructive moths) breeding in the walls and eating holes in her books and clothes. She hadn't found a nontoxic solution. If any of you have some tips you would like to share about this, please let us know so we can include them in another edition of this book. Thanks!

Planning for future additions

Planning things in advance always saves work, so think ahead. If you know you'll be adding on later, build the beginning of the foundation for the future addition where it attaches to the original foundation. This way the foundations will be tied together well. Put in the door frame that will access the new room. You can either hang the door or temporarily board up and insulate the opening.

As you cob the surface where the addition will one day be added, partially bury sticks and leave them sticking out into the future wall's location.

You can also leave keying holes in that part of the wall. Where walls are protected from the rain, you can leave a stair/step with sticks and holes.

Interior walls

Interior walls take up precious inside space. You may want to make them as thin as possible, unless they are load bearing and/ or sound barriers.

All the interior partition walls can be partially supported by the main cob walls at the places where they meet. Elsewhere, they can be supported with posts or milled lumber running from the ceiling to the floor. These posts can be buried into the floor or set on top of something to keep them off the ground. The wooden uprights can be used to help support the roof and/or a loft.

Wattle and daub is a name for a type of construction used in many countries for centuries. Walls are made of woven sticks to form a net or mesh, then covered on both sides with a thin layer of cob. This is a way to make thinner interior walls. Thinner walls inside are an advantage because you don't use as much precious space.

The woven sticks can be beautiful room dividers just by themselves.

Interior walls can be made using woven sticks, metal mesh or fencing covered with layers of newspaper or magazine pages dipped in clay slip. (See page 147 for more details.)

2x4 frame walls can be used for interior walls. Because these are so straight and flat, it takes a little imagination to get them to look right with the organic shapes of the cob. Plywood and sheet-rock will take an earthen plaster or thick earthen paint to help them match the cob walls.

Thin cob walls will block the noise pretty well, giving you more privacy between rooms. If the interior walls are not responsible for holding up the roof beams, they can be made as thin as 5 inches at the top with a 1 inch in 3 foot taper. Make some test walls and see how thin you can get away with.

If the interior walls are helping to hold up the roof or the loft, make them thick enough at the place where the roof beams will rest to support them.

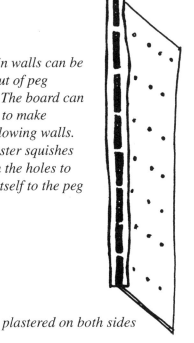

Very thin walls can be made out of peg board. The board can be bent to make lovely flowing walls. The plaster squishes through the holes to attach itself to the peg board.

plastered on both sides

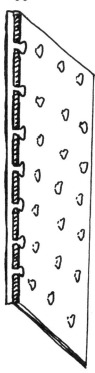

plastered on one side

101

TIPS FOR HAPPY COBBERS

Cobbing is most fun when done with others! Having regular weekend cob house-raising, potluck parties is a great way to get to know your neighbors, and a fun way to play with your friends and get your house built.

It is important for your health to **drink plenty of water** while you're cobbing. Because it doesn't seem like hard work, it's easy to get dehydrated. Encourage cobbers to drink by having inviting drinking water easily accessible on the site. Remind each other to drink.

Do like the English do and **have morning and afternoon tea** on the site. A little rest and a snack help keep cobbers keen.

Take time to **be artistic.** This is a wonderful opportunity for you to use all your amazing imagination and creativity. You can bury magic things in the walls. Make up rituals. Allow sacredness and beauty in your life.

Cob walls preserve things that are buried in them very well. It's fun to **bury a time capsule,** a bottle full of things that are likely to last a few centuries (coins, computer disks, ceramic, plastic). This will tell a story about our time in history. An archaeologist hundreds of years from now will have some clues about you, the mysterious modern/ancient builder.

Singing, drumming, and cobbing go together perfectly. Work goes faster and tiredness disappears! People can reach states of total ecstasy combining these three things. Remember to **enjoy yourself and your cobbing friends!**

It's fun to have a **special photo album/ journal** of the building process. This can be combined with a guest book, or have a separate book for co-creators and visitors to sign and make comments in.

If you want to **cob when it's too cold** for your fingers, wear those cloth gloves whose fingers have been dipped in green rubbery stuff. You can wear thinsulite liners under them for extra warmth. If you need to wear shoes, the less tread the better, because the mud won't stick to them so much.

The cob builders checklist

On the following page is the check list. Read over it to get ideas. This is to help you remember what to add to the walls as you build. I recommend that you enlarge, photocopy and laminate the checklist. Hang it up at the building site so it's easy for you to check it often as your home grows.

COB BUILDERS CHECKLIST

SELECT SITE
- [ ] develop access
- [ ] design house
- [ ] gather materials and tools
- [ ] get water to site
- [ ] set up temporary shelter for while you build
- [ ] plan drainage and floor levels
- [ ] create drainage
- [ ] do septic
- [ ] make cob tests

FOUNDATION
- [ ] dig for foundation to solid ground, below frost level— remember foundation for buttresses and fireplace (optional)
- [ ] plan for foundation insulation (optional)
- [ ] lay pipes for water and electric wires and for fire vent (slanting to outside of house & overhanging wall)
- [ ] build door threshold
- [ ] set up, square, plumb, and brace door frame
- [ ] put keying system on outside of door frame
- [ ] build foundation and door threshold

WALLS—FIRST TWO FEET
- [ ] put pipes in for water and electric wires if you haven't already put them in foundation
- [ ] set up electric
- [ ] start cobbing
- [ ] build in stove and/or fireplace as you go and bury woodstove pipe in wall (optional)
- [ ] start cantilevering for benches (top of seat height usually 14" from floor)
- [ ] put in low vent to admit air on cool side of house
- [ ] build in branches for future additions
- [ ] if you're having wooden upright supports for floor-to-ceiling shelving, put them up and brace
- [ ] incorporate wood into walls for (optional)
  - [ ] lower cupboard shelving
  - [ ] lower ladder rungs
  - [ ] bench support poles (inside & out)
  - [ ] opening for firewood box
- [ ] install electric outlets (optional)
- [ ] angle backs of benches so they're comfy
- [ ] roughly level floor(s) as you gather cob mix material
- [ ] establish wall taper angle as you build
- [ ] build window seats

WALLS—TWO TO FOUR FEET
- [ ] key in window sills and windows on sun-facing wall
- [ ] bury shelf supports or create cantilevered cob shelves
- [ ] sculpt in niches
- [ ] leave ledges to support counter-tops or bury counters into walls; if necessary, add posts for counter supports
- [ ] put in openable windows (keyed to cob)
- [ ] bury scaffolding supports 3-1/2 to 4 feet up walls (inside and out) and brace them
- [ ] keep incorporating branches for future additions
- [ ] continue wall taper

WALLS—FOUR FEET AND UP
- [ ] put in lintels or arches over windows
- [ ] sculpt art into walls or fatten wall for future carving art (optional)
- [ ] add more niches and shelves
- [ ] sculpt/install lighting fixtures
- [ ] incorporate wood for
  - [ ] cupboard supports
  - [ ] hanging artwork
  - [ ] branches for closet racks, hat & coat hooks
  - [ ] hooks for hanging kitchen stuff (pot racks)
  - [ ] curtain rod supports at tops of windows
- [ ] build in wooden support poles for porch roofs, awnings, firewood shed, etc.
- [ ] if you're making a loft, leave ledges and bury support beam in the wall
- [ ] put an openable vent on highest wall

ROOF
- [ ] bury rafters and keys for rafters
- [ ] put up roof beams and rafters—attach keys
- [ ] put up ceiling or leave a place in cob to slide ceiling sheathing onto
- [ ] insert screened tubes in cob between rafters to ventilate insulation space
- [ ] insulate roof well
- [ ] install roofing
- [ ] hang gutters (this is very important!)
- [ ] put in stove pipe, chimney, flashing, skylights

PLASTERING
- [ ] if you're using lime, mix up putty and let it sit at least two weeks
- [ ] make and analyze some plaster experiments
- [ ] carve and shape walls
- [ ] wet walls
- [ ] plaster (multiple layers if desired)
- [ ] embed tiles vertically along "wet" counters and for decoration on sills, etc.
- [ ] paint walls or lime wash (optional)
- [ ] preserve exposed wood

FINISHING TOUCHES
- [ ] hang doors and windows
- [ ] hook up plumbing fixtures
- [ ] doelectrical finishing work
- [ ] wood shelves, counters, shutters, lids, curtain rods, cupboard doors

FLOOR
- [ ] if you've put off doing drainage, do it now!
- [ ] do some floor material experiments
- [ ] roughly level floor if you haven't already
- [ ] build step edges at elevation changes
- [ ] tamp
- [ ] for an earthen floor, level guide nails; lay floor; dry; lay next layer; dry; oil, and/or wax

MOVE IN & BEAUTIFY
- [ ] create landscape elements
- [ ] porches
- [ ] outdoor walls
- [ ] paths
- [ ] trellises
- [ ] benches
- [ ] outdoor fire
- [ ] gardens

WINDOWS AND DOORS

The cob around the window and door openings supports the walls above.

You will need to make sure that the weight over the windows and doors is transferred to the columns of cob on the sides of these openings. The weight will then travel down through the cob and to the foundation below.

The columns of cob between the window and door openings need to be sturdy, and the space over openings needs to be spanned with a big hunk of wood (a lintel) or a cob arch.

When you are planning the placement of the windows and doors, make sure there will be **enough cob between openings to support itself and everything above it** - no weight should rest on the window or door. A cob section that's a foot across should be more than enough in most situations. Less will do for short walls (5 or 6 feet) and where there will be minimal roof weight.

Arches

Arches are very strong. The steeper the arch, the stronger it is. There has to be enough cob in the arch to support the weight of what's above it.

Arches can be used over plain glass or over framed windows. If you have a piece of glass in a frame that you don't want to see in the finished arch, simply bury the frame in the cob. You can arch over paned windows too. A clever way to arch over a rectangular door is to put a window on top of the door frame, and make a cob arch over the window. This is a good way to divert the weight over the door if you don't have a lintel.

To form the arch, sculpt it by hand a little bit at a time. Treat it like you do the rest of the cob. When you've finished cobbing for a while, make holes in the cob where you plan to add cob later. Let it harden enough to continue adding cob, making sure that it's remoistened and that the new cob is incorporated into the old. If you live in a dry climate, you can work on the arch first thing in the morning, let it dry awhile, and then add some more cob in the afternoon. You don't have to be too perfect - when the arch is completed, you can go back over it with your water bottle and big knife to give it a final shaping. When you do this, be careful of the glass! If you plan to plaster, remember that you will be adding the thickness of the plaster around the window.

With small windows (under 16 inches wide), you can be more creative with the window shape without using an arch or a lintel over the window.

Lintels

"Lintel" refers to the big hunk of wood that spans the top of window and door openings. Lintels are necessary to distribute the weight from above an opening to the sides of the openings. They are also sometimes called headers. Lintels are usually big hunks of milled or unmilled wood. A milled lintel can double as the top of the frame if you position it carefully.

When people cut large beams to size, they often save the off-cuts and then never find a use for them. It's surprising how easy it is to find pieces of big dimensional lumber just looking for a life.

If you have access to big hunks of metal or railroad tracks, I imagine these would make good lintels if you are short of wood.

Lintel thickness

Lintels need to be made out of sturdy, sound wood - at least 4 inches thick for milled lumber, and at least 6 inches for unmilled lumber. The bigger the better. The look of really huge pieces of wood suits the massiveness of cob walls. The thickness you'll need depends on the width of the opening and the weight that will be above it.

Lintel width

Because the lintel is holding up the wall over the opening, it will need to be about the same width as the wall. You can use one large piece of wood that's as wide as the wall, or several smaller pieces of wood laying next to each other to make up the width.

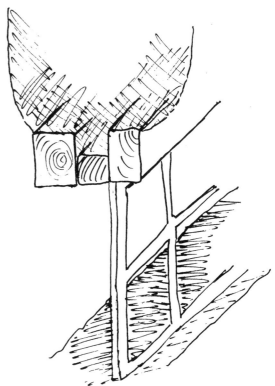

The lintel can be slightly narrower than the wall and the cob can bulge out above it. The lintel can be cobbed over on the outside and/or inside if you would rather not see it.

It's a little easier to get the cob to sit on squared milled lumber, but logs work OK and give a very ancient look to your home. If you choose to use logs for your lintels, you can fill the space between the lintel and the frame with cob.

Lintel length

Lintels need to be long enough to span the opening and rest on the columns on either side of the opening. The question often asked in the natural building circles is: how much do the ends of the lintel need to extend into the cob? This depends on a few things: the strength of the lintel, the size of the opening, and the weight the lintel will be supporting. I have never experimented to the point of failure, but have been surprised to find that as little as 5 or 6 inches of lintel buried into the cob works well. Use your own judgement and intuition.

Putting the windows and doors in the wall

See pages 36-37 for important tips on putting in the doors.

Make the bottom of the window sill and build up the cob about 6 to 12 inches (more for a tall or heavy window) on the sides of where the window will be. This will help hold the window once you set it up there. Mix up some cob and bring it close to where you're working.

Make the bottom of the window sill. Set the glass, framed window, or window frame in, and check the plumb and level. (Plumb means in alignment with gravity, or straight up and down.) Then secure the window with cob. Check it often while the cob is still pliable to ensure that the window is in the position you want. Pushing and massaging the cob can move it unintentionally.

When you are making a cob sill for a window, check to make sure the sill is level. Double check it by setting the level horizontally on the frame or glass, holding the window level, and stuffing cob under it to support it in the level position. When you're putting the window in the wall, make sure it is plumb too - unless, of course, you don't want it to be plumb. You can fix up the sill once the window is held in place.

Positioning the windows and doors in the width of the wall

Because the walls are so thick, you need to decide where to place the windows in the thickness of the wall.

If the window is big enough for a **window seat,** you have an opportunity to make a wonderful place to sit! Cantilever the cob out into the room to form the seat. You may want to raise the bottom of the window 6 to 18 inches higher than the seat itself to create a back rest. This also protects the glass a little and gets the window up higher on the wall.

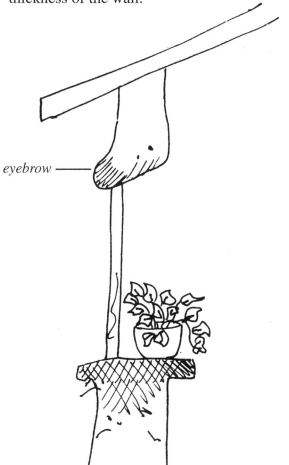

eyebrow

oops!

Remember that you will be tapering the walls, so you may want to make sure the top of the window or door will still be within the wall as it tapers.

Putting the windows close to the outside edge creates a big sill on the inside. This is a great shelf or seat, adding space in your house. You **may prefer to have a big shelf on the outside.**

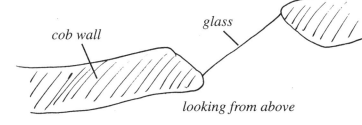

cob wall

glass

looking from above

The window can be placed in the opening at an angle to the wall for an unusual look.

You can put the window or door past the outside plane of the taper at the top, and fatten the wall above the window or door like an eyebrow. This is a good way to save the day if you forget to put the window or door within the wall's taper and gives the house an adorable hobbity look.

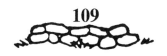

Plain glass windows set right into the cob

For the windows that don't open, simply embed glass into the cob. You can even use pieces of broken glass.

Remember to put tape around the sharp edges to protect precious cobbing hands. Bury the glass into the cob at least 1/2 an inch on all sides. The cob will shrink away from the glass as it dries.

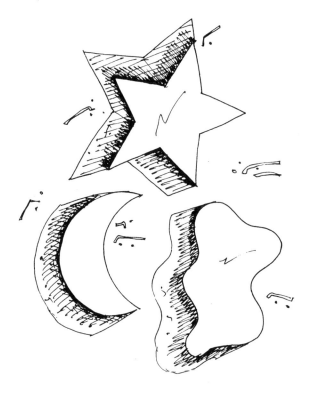

As you're cobbing, make sure to **check the plumb of the glass** with the level often. If you don't want the glass to make a bright reflection from a distance, angle it down toward the ground slightly.

For large windows you, must either sculpt an arch as the basic shape over the top of the window, or use a lintel to support the weight of the cob and possibly of the roof above the opening. You can shape the size and bottom of the window anyway you like. This is a wonderful artistic opportunity.

For small windows, (under 16 inches wide) you can be more creative with the window shape without using an arch or a lintel over the window.

Framed windows and doors

Ventilation is vital to a natural temperature control system, and fresh air is vital to people. Read the passive solar and ventilation section later in this chapter (page 114). Openable windows or doors are a way to ventilate your home. To open and close, they will need to hinge, pivot or slide in a wooden or metal frame embedded into the cob. Making the frames is basic carpentry. Ask a carpenter friend or read a carpentry book about installing opening windows in their frames if you need help with this part.

If you are using recycled windows, it's easiest to fix them up before you put them in the wall. Sanding, painting, replacing panes, and glazing are all easier to do when the window or door is laying horizontally on saw horses. This way your house will also look fancier while you are building, when curious folk come to see what that wild person is up to!

If you set a smooth wooden, vinyl, or aluminum window frame into a cob wall, there is a small chance that it could be pushed out of the wall. If the frame has a heavy door or window swinging on it, the frame could possibly work its way loose from the wall. To make sure the frame is well attached to the cob, add a strip of wood, 2x2 inches or thereabouts, running up and down along the sides of the frame. This is known as a **keying system** and will prevent the window from moving horizontally. Add 'rungs' to the strip on the frame to make it extra strong.

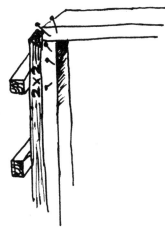

Use wood that is least likely to rot or get munched by termites. Find out which wood is the most long lasting in your area by asking local builders. Don't be too meticulous about your carpentry on the keying system because you won't see any of it once it's buried in the cob. Screwing the wood together is stronger than nailing it. To keep the frame square while you're cobbing, nail or screw a diagonal brace or two to the frame.

Another way to secure the frame in the cob is to put old nails or screws part way into the frame and bury them into the cob.

When the cob dries, it may shrink away slightly from the frame. That's OK. The space can be filled with plaster later.

Remember to check the plumb and the level on the frame when you set it in place, and as you cob it in.

The window and door openings are places that are more vulnerable to the weather. One way to help protect the window or door and its frame from rain is to build a cob eyebrow protruding out of the wall over the opening.

Another way to shield the window or door is to build in a mini-roof by burying the 'rafters' into the cob above the opening. The sheathing and shingles for the mini-roof can be added later. You can also keep the window drier by embedding wooden shingles or roofing tiles directly into the cob over the window.

The bottom of the window frames

When visiting old Native American ruins, I noticed that the wood that was buried into earthen walls centuries ago has survived the test of time amazingly well. As long as the walls are dry, the wooden window frames should hold up for a long while.

Wet causes wood to rot and makes it more inviting to bugs. The bottoms of the frames are the most likely part of the frame to get wet. Take extra care to keep them as dry as possible. Minimize the amount of wood on the bottom of the frame and bury it into the cob as little as possible. If it does get wet, it can dry out quickly where it's exposed to the air. Termites are a potential problem for wood buried in cob. (See the section on termites, page 100.)

The outside window sill

The cob on the bottom of windows is a place that might get wetter than the rest of the building. You may want to protect this area by setting large flat stones, a wooden sill, or bricks into the cob, forming a sill. These can be angled slightly downwards to the outside and overhanging the wall below. It serves as a tiny roof eave making the water drip away from the wall.

If you raise the wooden window frame up off the sill on a little cob ledge or on bricks or stones, the bottom of the frame will stay drier.

What to do about possible shrinking around the windows

Cob shrinks slightly as it dries. This means there is a chance that the windows could break under the shrinking pressure and weight. I have put many windows straight into the cob without doing anything for potential shrinkage and so far haven't had any problems. You may want to take the following simple measures to prevent any chance of undue stress on the windows and frames. Better safe than sorry.

Here are a few things you can do:

- **Let the cob on the sides of the openings dry quite a bit before building over the top of it.** Because building with cob is a slow process, the walls or cob columns beside the windows will usually be drying as you go, doing a lot of their shrinking before you get to the top of the window. If you are having a house raising party and building quickly, the columns may be wet from the bottom of the window to the top. Stop at this point and let the cob on the sides of the window dry for at least a week or two before going over the top of it. The taller the cob beside the window, the more it will shrink as it dries. If you have a high clay content in your cob mix, it will shrink more than a cob mix with a lot of sand and it will be more important to allow some settling space.

- **Be careful not to tweak the glass while you're building up the cob on the sides of the window.** While you are cobbing beside the windows, check the plumb and level often. Be aware that if you push the still wet cob near the glass you can distort the shape of the glass, making it more likely to crack as the cob dries.

- **Leave an air space for whatever spans the top of the window opening (a lintel or cob arch) to settle down into as the supporting sides shrink.** This will prevent the building's weight from crushing the window from above, if the cob columns shrink.

For lintels: You can leave a space between the window and the lintel. After the building has dried, you can come back and cover any gap that's left by stuffing it with insulation and nailing some small slats or wooden molding on the inside and outside. If you've let the supporting cob dry pretty well, and your mix has a high sand content, you probably won't need a settling space. If the supporting cob is still wet, a 1/4 inch air gap should be more than enough. Use your own judgement.

For arches: To create a settling space on top of a window glass or frame, tape a sponge (carpet underlay works great) or a small bundle of straw or old clothing onto the top of the window frame or glass. Bury this as you cob over the arch. The soft material will squish down if the cob settles.

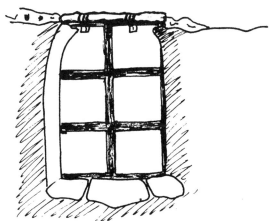

Things to consider before putting in the windows and doors

Passive solar design getting the most out of your windows

Read the section on passive solar design on page 12. There are lots of books available on passive solar design. Do more research to give yourself a more in-depth understanding. The principles of solar design will have very different applications depending on the climate and latitude of your home. Check with your local building department and natural building organizations for information that's specific to your area. I will give you a very simplified version here that applies mainly to a temperate climate.

Glass is magic stuff. It lets the sun's radiation through to heat up the surfaces in the house. Materials that are heavy and dense, like cob or stone are good thermal mass. They absorb the heat and radiate it back. Place your windows so a lot of sun will shine directly onto the floor. (See page 58 for more about this.)

In the winter when the sun is closer to the horizon, it will shine deeper into your house. This will help heat up the floor and the cob during the winter months. Your roof overhang will protect your windows from the sun as it gets higher in the sky during the summer.

Set your windows into the wall in such a way that they will get the winter sun and be protected from the summer sun.   Run the house length east/west if you live where you want maximum exposure to the sun, or angle it up to 20° to the east for morning light.

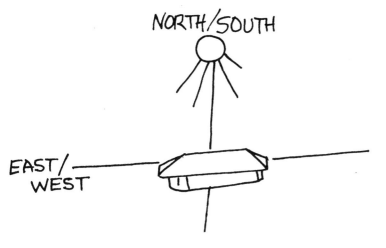

NORTH/SOUTH

EAST/WEST

Here's a flexible rule of thumb for a temperate climate. Put roughly 70% of your total glass facing the sun at its highest point in the sky. (This will be south in the northern hemisphere and north in the southern hemisphere.) Put 20% to the east to welcome the sun and start your home heating up quickly after the cool night. Put about 5% to the cold side of the house and 5% to the west where you probably won't want the lowering summer sun shining in after it's been heating up your house all day. Check out your weather and adjust this formula to suit it.

Glass is not very insulative and will let some of the heat and cold in and out of your home. Too much glass will make for extreme indoor temperatures. The local building department will be able to tell you the ideal amount of glass for passive solar heating in your area. This will be a percentage of the floor area, for example: put in as much glass in the wall facing the sun as it takes to equal 15% of your floor's square footage.

Different types of glass have different insulating abilities. Call your local window store to learn more and check out the current prices. If you live in a temperate climate, it is required by code to install double-paned vacuum-sealed windows with at least an R4 insulating value. If you buy double-paned windows, I recommend buying from a reputable company. The windows should be guaranteed for life, but they do leak and fog up sometimes.

What about making your own "double-paned" windows? In the old days, people put up storm windows during the seasons of extreme temperatures. These are a separate single pane in a frame that can be removed for cleaning. Unless they're vacuum-sealed, condensation will form between the sheets of glass. Make them so you can remove one of the pieces of glass to clean the insides in case they fog up and grow algae. The insulation value of storm windows is not as good as vacuum-sealed, but the better the seal, the better the insulating ability. Because **single pane windows are free** or cheap, you may choose to use them if you don't want to buy double-paned windows and if you're not worried about code requirements.

The simplest passive indoor-climate-control system is to make openable windows and open them and the doors when the outside temperature is what you like, and to close them all when it's too hot or too cold, trapping the temperature you want indoors. This is one of the most effective ways of controlling indoor temperature, especially if the temperature varies a lot during the night/day cycle.

Covering the windows with insulating curtains, shutters, or clear plastic helps keep some of the cold from flowing in and the heat from flowing out. If you want **to keep the sun's heat out in the summer**, shade the windows on the outside so the sun doesn't get in to heat up the indoors. Using a reflective material works best. Hooks to hold up the shades and curtain rods can be built right into the cob, inside and out.

Block off the small gaps around the windows and doors where the heat or cold can get in and out. To make a **natural caulking, dip a string in wax** (beeswax is more flexible) and push it into the space you want to fill. Strips of inner tubes can also be used to seal cracks. Bigger spaces can be filled with cob.

Another advantage to passive solar design is that nature's light is cheery and good for your eyes. **To get the most light out of a window, put it high in the wall** unless it's blocked by the eaves or a tree. Place the windows so they let light into the places you want it, like the desk, the kitchen sink, counters, etc.

There's nothing quite as nice as natural light from above. Skylights let in more natural light than wall windows. There are some disadvantages though. Since heat rises, a skylight will let out the heat on a winter's day and may overheat the house on a hot summer day. You can splurge and get one that has double or triple layers of glass for insulation. Plan to cover skylights to keep the summer heat out and/or make them openable to let the heat flow out. They require careful attention when installing or they'll leak.

How to get lots of glass in the sunny side of the house and still support what's above

You will need columns of cob that are at least 10 inches thick between windows and doors to support your structure. On the sunny side of the house, these columns will block a lot of the sun that you may want for warming your home in the winter.

One way to increase the sun that shines through is to use large windows and angle or bevel the supporting cob-columns away from the glass. This lets in more light when the sun is shining at an angle to the window.

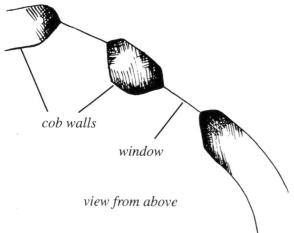

cob walls

window

view from above

Another way to maximize solar gain is to put two large windows side by side with a wooden board between them and one long lintel over the top of both windows, resting on the cob at either side.

Bay windows are lovely and easy to make. Build the foundation in the shape of the bay you want. You can make cob pillars between the windows to support the arches or lintels above them. You can use pieces of strong wood between the windows as posts for the lintels to rest on. Bay windows are lovely places for window seats.

If you want to have all glass on the sunniest wall, use a post and beam framework instead of cob between the windows, and all the way up to the roof on that wall. Get an imaginative carpenter friend or carpentry book to help if you need to. Where the wood wall and the cob walls meet, use a heavy duty version of the window frame keying system, or use a box key.

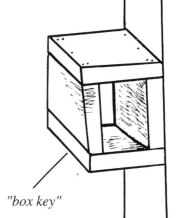

"box key"

Trombe walls

A trombe wall is a cob wall with a glass wall in front, to trap the heat. You can build a glass wall in front of a sun-facing cob wall anywhere from 4 inches away,, to far enough away to serve as a hallway or sun-room. Close off the sides between the glass wall and the house wall with cob or wood. The low winter sun's rays will penetrate the glass and heat the cob wall during the day, which will radiate the heat into the interior space at night. A dark colored wall will heat up more quickly than a light colored one.

Another option is to put vents on the top and bottom of the house wall to circulate the super-heated air from between the wall and the glass into the interior of your home. With this system, you'll need to set up back flow dampers to stop the warm air from inside the house flowing back into the trombe system.

If your large trombe wall will cover the sunny side of the house, it's OK to put windows in the cob sides of the trombe system so you can see out and enjoy the natural light. If you make the windows operable, opening in, you can clean the glass on the insides of the trombe wall and outsides of the windows. Set up a shade or plant deciduous trees, artichokes, tomatoes, or something to protect the wall from the unwanted summer sun's heat.

You can find out more about trombe walls in books on passive solar design.

Ventilation

Observe the sunshine, wind and the climate carefully at your site before deciding where to put the windows.

Fresh air is delicious! Openable windows will allow fresh air into your space. When deciding where to place openable windows and doors, think about how air moves. Notice which direction the wind usually comes from if you want cross ventilation. Heat rises, cold sinks. For summer cooling, put a vent high in the wall to let heated air out, and one low on the cold side to let the cooler air in to replace it.

A very simple vent can be made out of a bucket with its end cut off. Any sort of tube buried in the wall will serve the same purpose. These can have removable or permanent screens attached to the outside of the vent to keep critters out. I've seen bright colored cloth tied onto the outside end of the tube or bucket to give a pretty light and keep the mosquitoes out. When the vent needs to be blocked off, put the lid on the bucket or just stuff a blanket or something in it. These buckets can second as storage places.

You may want screens on the window openings to keep insects out of your house. I like to have removable ones because I don't need them in the winter, and I love to look outside without anything interrupting my vision.

Make extra ventilation in the 'moist' rooms like the kitchen and the bathroom. This will help keep them drier. You may want to put a vent over the stove to let out cooking smells and grease.

Make good latches to secure your windows and doors, so the wind doesn't fling them around when they're open, and so they won't rattle or leak when they're closed.

In windy areas your house will interrupt the wind flow, causing the wind to whirl around on the lee (downwind) side of the house. Positioning openings to avoid thewind blown dust can be tricky. Talking to the neighbors about how well their house designs worked could be helpful.

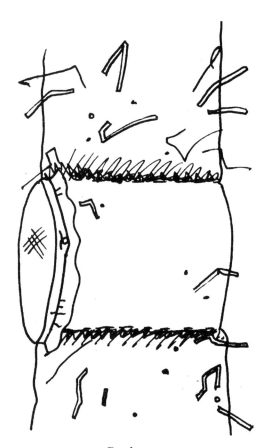

Bucket vent

Views

Imagine your life in the house as thoroughly as you can. Where will you sit to eat and to study? Where do you stand to do the dishes? **What view do you want to look at** while doing the things you do in your home? Where will you need natural light? What do you want to see from your bed?

Remember, windows allow people to see into the house as well as out. When visitors approach the house, what do you want them to see? If you have neighbors close by, think about where their windows are and how you want your windows to relate. Where do you want privacy?

When you've built the walls up to the height of the bottom of the window, have a friend hold the window up in different places so you can see where you like it best. Because it's so exciting to put windows into the walls, the tendency is to put them in too low in the wall. Observe window heights in finished homes to help you 'see' what you want. Stand and sit in the house while your friends are holding up the windows to check if the height is right. Consider this carefully before you place the windows.

Try to imagine the finished building. Look at it from the outside as well as the inside before you make your decision. You may want to get more than one friend to hold up other windows, so you can get a look at how they all relate to each other.

Noise

Cob walls block sound amazingly well. Windows will let noise in and out a lot more than the walls will. If you live near something noisy, you may want to minimize the windows facing the direction of the noise.

Magic windows

Centuries ago, people sometimes used window placement to mark a certain day of the year. You can put a window where the sun will shine through it on your birthday, a solstice, or on the equinoxcs, and onto a special spot on the opposite wall. This is tricky to figure out, unless you can capture the moment as you're building. Anybody know how the people in the old days figured it out?

Some glass safety tips

If you can get it, use safety glass in the most vulnerable places: skylights, large windows, and doors. (Safety glass is the kind used in cars, that's laminated so that it breaks into small pebbly bits instead of big dangerous shards.) Transport and store sheet glass in an upright position. Put it in a safe place, where the wind or someone walking by won't knock it over.

Have a lot of respect for the danger of being cut with glass. Wear leather gloves and shoes when handling it. Never carry glass around after drinking alcohol!

Be extra careful when handling or storing broken glass. If you have broken glass that's still hanging onto its frame, it may be safer to take it out of the frame before moving it. When you are embedding broken glass in cob, put tape around the sharp edges and wear gloves. Be extra aware of your precious hands while you are massaging the cob onto the wall around the glass. Cobbing with cuts is not so fun.

When you put big pieces of glass into the walls, put some long strips of bright colored tape onto the glass to prevent yourself and others from accidentally trying to walk through it.

119

Getting rid of unwanted windows

Too many windows is easier to deal with than too few, so be generous. Later you can cob over any window you want to get rid of. It's harder to knock out a piece of the wall and add a window than it is to cob over one.

Replacing broken glass

If one of the windows that's buried in the cob breaks, you'll have to chip the cob away, get a new piece of glass, and cob it in again. Cob sticks to cob well. Read the sections on applying new cob to dry cob (pages 91).

Fun window ideas

Bottles for windows

Bottles and jars can be buried into the wall to make fun mini-windows. They are more insulating than a single pane window because they already have two "panes" or sides. Colored ones will light up with color when the sun shines through them. When you're outdoors at night, the light from inside will look lovely where it shines through the colored glass.

If the parts of the bottles and jars that are buried in the cob are white, they reflect a lot more light than if they're right up against the dark cob. The outsides can be painted white, or wrapped with wide white tape where they will be touching the cob. Leave the exposed parts uncovered so the light can shine through the bottles.

If you bury a tapered bottle (like a wine bottle) into the wall, the cob around the skinny end will block most of the light. You can slide a jar over the small end and cob that in. This will allow a lot more light through.

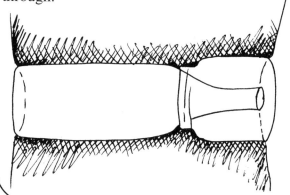

With a little imagination and a lot of colored bottles, pieces of glass, ashtrays, and stuff from second hand stores, you can make amazing stained glass/cob "windows". Bottles and jars can be cobbed in sideways to the wall too, tapering the width of the cob wall in toward the glass.

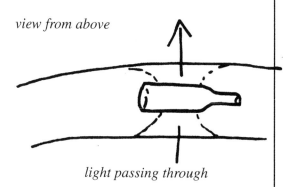

view from above

light passing through

Big gallon-sized jars and buckets can be buried with their mouths open to the inside or outside of the house to be used for storage. You can use their lids to close them.

Glass from cars

Windshields and the rear windows from cars are often cheap or free and make interesting windows buried into cob. All car windows are made out of nice thick safety glass. You could even bury a whole car door into a cob wall, leaving the window handle and the window exposed for an openable window.

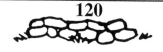

ROOFS

There's lots of information about roofs out there, so I will give a very basic explanation and tell you about the eccentricities of attaching a roof to a cob building. I'll also briefly describe some 'out of the ordinary' roof systems.

The roof is usually the part of your home that costs money. This often means milling, mining, toxins, transporting, etc. You may want to consider the environmental impact of your materials and use recycled stuff whenever possible.

The roof is absolutely vital to the longevity of the walls in any type of home and is often the weak point, requiring maintenance and rebuilding. The roof passes the weight down to whatever is below, then onto the walls, down to the foundation and the ground. It is worth putting time, attention and good quality materials into your precious roof.

A vital part of designing the roof is to decide where the water from the roof will end up. Think about this very carefully!

Components of a roof system

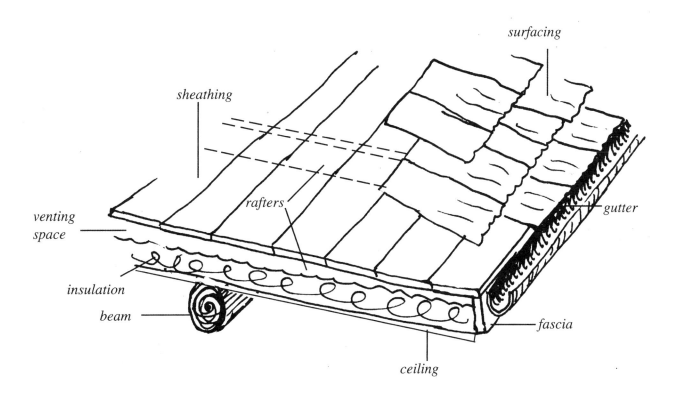

surfacing

sheathing

rafters

gutter

venting space

insulation

beam

fascia

ceiling

Beams

Beams are the big pieces of roof (wood or steel) that support the rafters, the rest of the roof and their own weight.

Beams can span the building or lay along the walls at the top and bottom of the roof slope.

They can be left out entirely in a small house, if you have rafters big enough to span from wall to wall.

Rafters with no beams

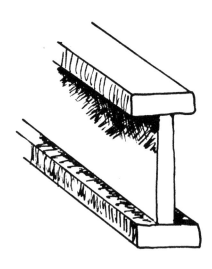

To save wood use I-beams
I-beams can be bought commercially or made yourself.

124

Rafters

Rafters slope the same way as the roof. The roof sheathing is often attached to them. Rafters sit either directly on the walls or on beams. One of their jobs is to support their own weight and the weight of whatever is on top of them. The rafters need to be close enough together to prevent the sheathing from sagging between them.

Rafters often form the sides of the space for insulation, establishing the height of that space. It's a good idea to have lots of insulation, so the rafters should be as deep as possible if they are to house the insulation.

Sometimes the ceiling is screwed to the rafters. If the ceiling is attached from below, when you are on the roof, be sure to **keep your weight on the rafters, not on the ceiling.** It could pull away from the rafters, and you may end up on the floor unexpectedly!

If you use unmilled poles for rafters and you're using a fairly rigid rectangular sheathing and/or ceiling (plywood or sheet rock), do your best to choose straight poles for rafters and try to line them up so the sheathing pieces can be nailed to the poles.

Boards can be attached to the ends of the rafters (called fascia boards). These boards can be used for the following purposes:
- to look good
- to protect the ends of the rafters
- to hang the gutters on where the rain runs off the roof
- if the outer rafters do not rest on the beams or cob, they can be attached to the facia board to hold them up.

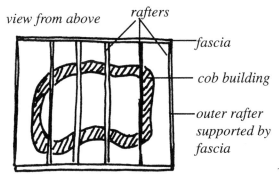

view from above   *rafters*

fascia

cob building

outer rafter supported by fascia

Nogs or blocks

These are short pieces (usually wood of the same dimension as the rafters), placed between the rafters to make sure they stay an even distance from each other and to help prevent them from twisting. The nogs are positioned so they can also provide a nailing surface and support for the edges of the sheathing.

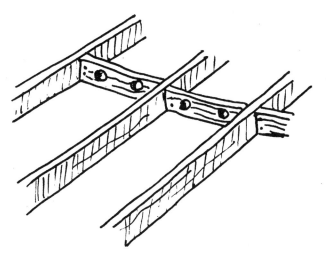

If the nogs are used in the insulation space they must have large holes drilled in them to allow for ventilation of the insulation. Drill the holes before you put them up.

Bracing

Cross bracing makes any angle many, many times stronger. If you plan to extend a beam or rafter out past the wall and you want extra support for the overhang, you can think ahead and bury a diagonal brace into the cob.

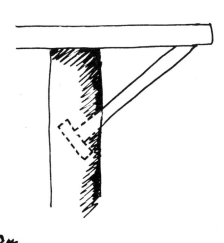

Vertical Posts

If you are using posts to support your roof before you build your walls, it is important to design them so they will be outside (or inside) the cob walls, but not surrounded with cob. If an upright post is buried in a wall, it will create a weak point where cracks will form as the cob settles and dries. (See illustration page 134.) You can bury posts into a low wall (2 or 3 feet) and the cob will be OK. It is very important to have good drainage under the partially cobbed posts. If you live where termites are a problem, it might be best to keep buried wood to a minimum.

Take care to protect the bottom of the posts from moisture and support them up off the ground. If you plan to bury your posts in the ground, you can make them less inviting to rot and bugs by charring the end in a fire.

If you set the post back from the very end of the roof beam it will stay drier, and will also make it a little trickier for critters to get up onto your roof. A diagonal brace from the post to the beam will help secure the connection between the two and encourage the post to stay plumb.

If you are supporting the beams on posts, it is strongest to set the beams directly on top of a post - rather than attaching them to the side of the post and depending on bolts or big nails to keep them there.

Roof sheathing

Roof sheathing refers to whatever you use on top of the rafters to support the roof surfacing. Plywood, chip board, and wooden boards are the most common choices of sheathing materials. Tongue and groove boards are often used. (See page 136 for more on roof sheathing.)

The sheathing adds strength and diagonal bracing to the roof structure. There is usually a ventilation space below the sheathing and on top of the insulation.

The sheathing can be placed flush with the ends of the rafters and the fascia can butt up to the bottom edges of the sheathing.

Gutters

Purposes of gutters:

- To stop the water from pouring off the roof and splashing up onto the walls. This is a common cause of erosion at the outside base of the walls in old cob homes.
- To keep the runoff from the roof from pouring onto the ground next to the house and seeping under the foundation, possibly causing uneven settling.
- To collect water that falls on the roof and direct it into the down pipes and then to somewhere useful.

Put the gutters on sooner rather than later. If you procrastinate on this job, you will put the health of your home at risk!

You would be surprised how much water is collected on the roof. Make the gutters big enough to handle a large amount of water and think carefully about what happens to that water. Water is a precious resource if it's where you want it, but it can also cause a lot of erosion as it flows out the down pipes.

It is very important to **extend the roof surfacing and/or sheathing out over the gutter** at least 3/4 of an inch.

The water collected by the roof should fall into the gutter - not down between the gutter and the building.

The gutters must slope toward the down pipes at least a 1/4 inch for every 20 foot length of gutter to make sure the water flows along it.

If you have an organic shaped roof, it is very tricky to get the gutters to follow the edges of the roof. **Remember to design your roof with the gutters in mind.** Check out what your local gutter people have available and ask them for any suggestions. You might want to try making your own gutters out of wood and covering them with painted cloth. (See page 140 for more on this idea.) You could also experiment with gutters made of bamboo, large diameter pipe cut in half, tile, etc.

There will be more on roof surfaces, insulation and ceilings later in this chapter.

Some common roofs

You can use any of the following types of roofs or use more than one kind in combination with each other.

Domes and vaults

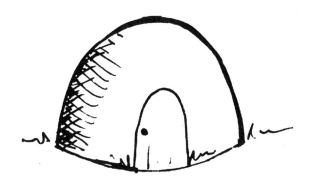

In many countries, buildings are roofed with domes and vaults made out of unbaked earth. I have never built a domed or vaulted cob roof on a full sized building and I don't know anyone who has. I imagine that, cobbing carefully, it could be done.

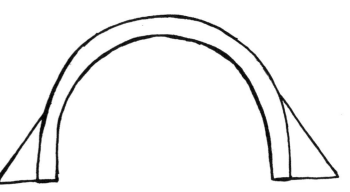

The weight of the domed or vaulted cob roof puts a large amount of outward pressure on the walls below them. The bottoms of the walls may need to be built extra wide or be reinforced with buttresses.

Earthen roofs are commonly built in climates with very little rainfall. In wetter areas, a water-repellent coating would be needed to protect the earthen roof from eroding or from becoming saturated with water and possibly collapsing. The earthen roof would also need to 'breathe' out interior moisture through this roof coating. The natural building world will be delighted when somebody perfects such a material. Making buildings entirely out of earth, using very little wood or manufactured materials is very exciting. (See the Plaster Chapter, note Adobe Protector, page 166.)

Any of the following roof type structures can be made with peeled poles, milled lumber, bamboo, or steel weight bearing members.

Cone shaped roof

This shape looks right on a rounded building. A cone roof shares its weight evenly around the tops of the walls. The junction of the rafters in the middle of the roof may seem a little daunting...

Some options:

Bolt the rafters to something rigid like steel or plywood.

Bolt or screw the rafters to each other at the top of the roof.

A tension ring, further down the rafter, takes a lot of the stress away from where the rafters meet at the top and minimizes any outward pressure on the walls. A simple tension ring can be made by running a cable through holes drilled through the rafters, or by attaching a cable to notches in the ends of the rafters.

A thatched cone-shaped roof is very common on many African cob homes.

The rafter ends will run out over the walls, creating the eaves of the roof. The sheathing and the ceiling pieces need to be cut at angles to cover the spaces between the rafters. If you put the rafters the same distance apart from each other, the angles should end up the same. If you arc thatching this roof, you can weave a basket-like frame to tie the thatch to.

Shed roof

A shed-type roof is one plane, the easiest kind of conventional roof to make. If you've never built a roof before and you don't have a dedicated carpenter friend, a shed roof might be a good roof to choose.

Double shed with clearstory windows

This simply means two shed roofs at different heights. This can be a very practical design for passive solar and natural light. The glass between the two roofs can be placed to let the sun shine deep into the cool, dark side of the house.

Gable roof

A gable roof is a common design with two roof planes meeting at a peak or ridge. This style of roof gives you the option to have an attic. Attics can be used as spare rooms, for storage, or for bulky insulation like whole straw bales.

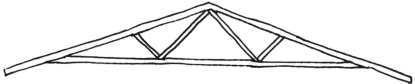

You can save yourself time and effort by buying sets of prefabricated rafters with bracing for gable roofs. These are called trusses. You can also build them yourself, either on the ground or in place on the roof. Go to the local lumber supply store and learn more about trusses, what's available, and their prices.

A gable roof with a shed roof on one or more sides looks great and is a good design for a home that is built in stages.

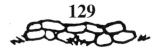

Gambrel roof

For a loft or two-story home you may want to put another angle in the gable roof design. This is called a gambrel roof. With the added steepness on the sides of this roof design, the walls can be shorter and you'll still have enough headroom for an upper story.

Hipped or Pyramid

This type of roof is a little bit trickier because of the angles that need to be figured out and cut. Books on roofing will make it easier for you. These roofs are especially popular for straw-bale buildings because they distribute the roof's weight to all the walls.

Organic shaped roof

The organic shapes of many cob homes lend themselves to flowing, organic roof shapes. These are more challenging to most folks than roofs that are on one or two straight planes. If you love the organic look and have a flexible mind and the time, go for it!

Nature will provide you with a supply of gracefully twisting and curving logs and branches to inspire adventurous roofs. You will need some sort of sheathing and roof surfacing material that will flex and bend to the organic shape. The small branches that thatch is attached to are perfect for making an organic shape, as is thatch itself. You could also sheath the roof with small sections of plywood or boards cut to fit from rafter to rafter.

For a sheathing that will flex to an organic shape, have a tree or two milled into long thin 1/2 inch boards. The boards will flex easier if they are put on while the wood is still green. The edges of these boards can be milled or left in the natural shape of the tree trunk.

Starting at the bottom, bend the boards to the shape of the rafters and screw them down to the rafters. The boards can be butted up to each other or laid with each board overlapping the lower one an inch and a half or more.

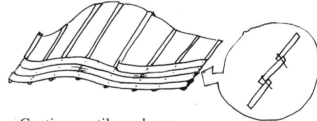

Continue until you have finished the sheathing.

Roof design and planning

Read over this whole chapter once or twice. Go to the library and find books on the subject. **The simpler the roof, the simpler the roofing job.**

The roof design is a big part of the external character of the home. Notice which roofs appeal to you.

When designing your roof structure, **consider the weight** of the roof components. The heavier they are, the stronger the supporting roof structure needs to be.

A common cause of failure in old cob buildings is the roof's weight pushing sideways on the walls and causing them to bow or crack. The steeper the unbraced roof, the more it pushes on the walls.

Supporting the roof laterally to itself will solve this potential problem. With bracing, the roof will become one unit with its weight going straight down onto the walls.

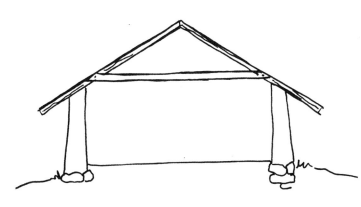

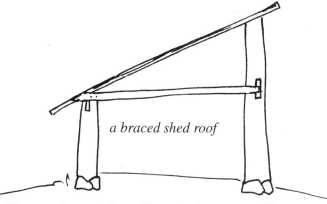

a braced shed roof

The smaller the building, the less you will have to concern yourself with bracing. There will be a lot less weight involved and a small building with thick cob walls is naturally sturdier. With a small building and a shed roof, if the supporting roof rafters have to span less than about 15 feet, it's OK to rest the rafters on the cob without bracing.

The distance a rafter or beam can safely span depends on the dimensions of the wood and the weight of the roof. You can check builders' charts to see how big a piece of wood you'll need to span the distance between your walls.

The steeper the roof, the faster the water will run off. The required steepness, or pitch, of the roof depends on the surface material. Some are better than others at getting the water to run off. Conventional roofing will come with a minimum pitch recommendation: this is often 3 in 12 or 4 in 12. These numbers mean that for every 12 feet the roof spans horizontally, it must rise 3 or 4 feet.

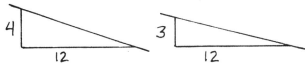

Make sure that every part of the roof is angled enough for the water to run off. The flatter the roof, the more likely it is to have leaking problems.

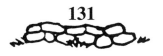

Dips or valleys in the roof are potential trouble spots. A valley is where the bottoms of two roof slopes meet each other on an angle. Water flows into the valleys from two directions, meets there, then a large amount of water flows down the valley. In snowy climates the snow gathers in the valleys, sometimes freezing, thawing, and refreezing there. This is a common place for leaks to develop, so take extra care when building the valleys in your roof. Debris also tends to gather there and will need to be swept off once in a while.

When designing your roof structure, plan where the insulation will go. The insulation can lay within the roof structure or on top of it, as with a thatch roof. In a design with an attic, the insulation usually sits on top of the ceiling, separate from the actual roof. An attic or large insulation space is necessary for bulky insulation.

insulation

If you make an open-ceiling type of roof (no attic), the depth of the rafters makes a space for the insulation and the ventilation. The ceiling must be well attached to the rafters and sturdy enough to hold up the insulation.

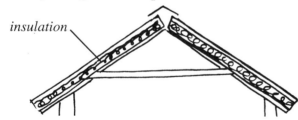

insulation

If you live in the tropics you can **let the heat out of the house by designing big adjustable vents** in the roof.

The roof structure will be the part of your cob home most susceptible to fire. Some roofing and insulating materials are more flammable than others, so consider the fire danger in your area when choosing your roof materials.

If you plan to collect the sun's heat with **solar panels**, the roof is a good place to mount them. Check in a book about solar energy to find the optimum angle for collecting the sunlight at your latitude. You may want to make the roof, or part of the roof, at that angle to set the panels on.

If you are building in a place where it snows a lot, see what you can learn about designing in snow country from the library and from local practice. Observe different roofs in heavy snow. **Think very carefully about where the collected snow will end up when it slides off the roof.** It would be very depressing if it piled up in front of your sunny windows or in front of your door.

Snow on the roof is added insulation but **it can also add a lot of weight to your roof structure.** If you want the snow to stay on the roof for insulation, build a sturdy enough structure to support it. If you want the snow to slide off the roof quickly, use a more steeply pitched roof and a smooth metal roof surface.

Porches and sheds add wonderful living and storage space to your dwelling. Every house needs at least one. For shed or porch roofs that run off the main building, you can put the rafters right into the cob walls as you build. Support the low ends of the rafters on a post and beam structure.

Another way to make outdoor covered areas around your home is to use extra long roof beams. Support them at the ends with posts, and have more rafters running between the extended beams.

If you live in a wet climate, it's a good idea to put gutters on these little roofs too.

The place **where the chimney, stovepipe, or skylight comes through the roof** is a favorite spot for naughty leaks. If these potential trouble spots are placed high up on the roof's slope, they will have less flowing water to contend with. The flatter the roof, the more likely it will be that you'll have leaking problems. **Be extra meticulous about sealing these areas.** A good carpentry book will give you some helpful tips.

If you choose to do a wooden roof structure, it will require basic carpentry skills. This is a good time to call on your carpenter friends.

If you plan to add onto your home in the future, it is important to design the roofs so they complement each other well. Design and build the first roof with the future addition(s) in mind. Make sure the water will run off every roof and not onto another roof. You may choose to build the whole roof for the finished house first and build the cob additions up to the roof as you find time. (See the next section for more on this idea.)

It's easier and more fun to roof your home with a group of people: some folks handing stuff up, three or four people on the roof aligning the beams or rafters and attaching the roofing. Watch out that the people on the roof don't drop things on the people below. **Be careful when you're on the roof.** Steep roofs are harder and more dangerous to work on. Gravity wants to bring you back to earth. Get up and down carefully. It's a good idea to stop and get off the roof before you get too tired. If you're working on a really steep roof, you can tie yourself up there with a safety rope.

Building the roof before the walls

You can build the roof before the walls and even before the foundation. There are some advantages to this, as follows:

You and the cob will have great protection from the sun and rain while you are making the rest of the house.

Putting up a pole-supported roof can be done with fairly standard carpentry skills. If you are pressed for time or energy and want to hire someone, it's much easier to find someone who knows how to put up a pole structure than to find someone who knows how to put a roof on a cob building.

Because wood is rigid and cob is malleable, it's easier to mold the cob to the roof than to make the roof fit onto the organic shape of the cob walls. Cob can fill all the odd little gaps between the walls and the roof. It's also easier to get the **ceiling** up first and then cob to it rather than trying to cut the ceiling into the organic shape of the walls.

Many counties will be willing to give you a permit for a pole structure, especially for an 'agricultural' building. They're often unconcerned about what kind of material you want to use to fill in between the posts. Currently, most building officials seem to know very little about cob.

You can support the roof on permanent posts or poles, **or with temporary posts** that can be knocked out after the cob has been built to hold the roof up. Cob walls are load-bearing and will hold the roof's weight. Permanent posts can be used outside the walls as porch supports, or inside as part of interior walls, or for beam supports. Plan the **posts** so they **won't be completely buried in the cob walls** - that would create a weak point in the cob. **Posts can be partially buried** in the cob. I have buried up to 2 and 1/2 feet of the base of a post, making the cob extra thick at that point, and it hasn't made the cob crack. **The cob wall can wrap around the posts, half covering them vertically.** Make sure the cob is approximately as thick as the rest of the wall.

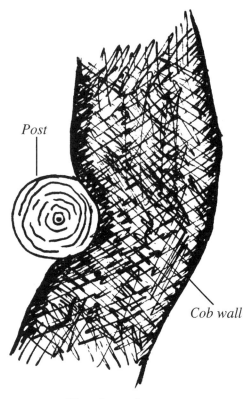

Post

Cob wall

View from above

Putting the roof on as you build the top of the wall

If you haven't built your roof first, don't finish the walls and then try to put the roof on them. You'll need to finish the walls and put the roof structure up at the same time. It is much easier to do these two things together.

The weight-bearing roof beams and/or rafters can be buried into the tops of the cob walls. Like the window frames, you can make a keying system to attach them to the cob. Another way to secure the roof to the building is to bury a piece of wood part way between the foundation and the roof and connect the wood to the roof structure with cable or wire.

cob wall

Wire

Wood

When you add the pieces of wood that connect the beams and/or rafters to the cob, think about the forces working on the junction of the wall and the roof. Each part of the roof must be securely attached to the next, especially in high wind areas. **The roof is like a wing - wind will try to lift it off the building and up into the air.** Gravity will try to slide it off the structure.

Embedding rafters

You can embed the rafters straight into the cob walls. If all the rafters are aligned with each other, it makes it easier for the sheathing to sit on the rafters and line-up for nailing. Measure the distance between the rafters, at the top and the bottom of the rafters, making sure it is the same before securing the rafters in place with the cob.

If your house is an irregular, organic shape and you're putting a flat planed roof on it, **making sure the rafters are in the same plane can be very tricky. You can align them with guide strings or set them on top of guide poles.**

Build the walls to the approximate height of where the rafters will sit. Stretch two or more strings or poles across the building, perpendicular to the rafters. Check that each string or pole is level. You can support the strings or guide poles on cob columns built on the wall, (these will later become part of the walls) or on temporary posts. Hold each rafter up where it will go using the strings or poles as a guide. This will show you how high you'll need to build up the supporting walls to rest the rafters in the same plane as each other.

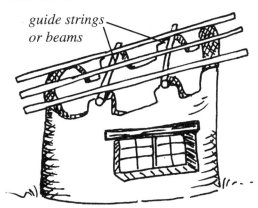

guide strings or beams

If you use guide poles, they can be left in to support the rafters like beams. If it's such a short span that the rafters don't need the support of these pole beams, they can be left in for looks or removed later by sawing them off and plastering over the stubs.

If you are using unmilled poles for the rafters, choose fairly uniform ones. This will make it easier to attach sheathing to the tops of the rafters and the ceiling to the bottoms of them.

To make an organic shaped roof, you can build the walls up to the shape you want, or use guide poles that are curved to rest the rafters on.

The rafters can also be curved or they can be straight to make it easy for you to put rigid sheathing on them.

The **rafters will need to line up with the edges of your sheathing and ceiling,** so take the time to measure them and place them accordingly.

Embedding beams

Permanent beams can be buried in along the top of the walls to support the rafters. These can be used instead of guide strings to establish where the rafters go.

It's ok if the walls are curved, and the beams straight, as long as the wall supports the beam well. Where the wall doesn't line up with the beam, you can simply cob up to the rafters.

If the rafters need more support, bury another beam or two across the house for the rafters to sit on. If you want a parallel look, make sure these beams are parallel to each other. If you want a flat plane on the roof, check that the beams are set level on the wall before cobbing them in.

Your beams can extend out past the walls to help support the rafters that sit on the outside and protect the walls of the house. The beams need to be big and strong enough to do the job. Diagonal bracing may be necessary.

How to get your big heavy beams up onto the tops of the walls

When you're about to move a big beam, prepare the spot where it's to be placed before you pick it up! If you will be bracing it, have the necessary bracing materials and tools handy.

Probably the easiest way to get a heavy beam up there is to get a bunch of friends together and lift it up. You may want to do this in two steps: lift it first onto the scaffolding or loft, then up onto the top of the wall. If you don't have enough people-power around, you can set up a ramp by leaning poles on the walls and pull the heavy beam up onto the building with ropes. Be careful!

Roof surfacing

The roof surface is the top layer of water-repelling or waterproofing material.

Many modern roofing surfaces - like plastic, tin, or asphalt - are moisture barriers. They protect against moisture from the outside but tend to trap the moisture created inside the house. Moisture barriers are often a place where vapor in the air condenses and problems arise. This potential problem can be reduced by supplying lots of fresh air to the area between the sheathing and the insulation. There's more about how to vent this area in the section on insulation. (See page 132.)

When attaching the roof surfacing to the roof, always start at the bottom of the roof and work your way up.

At the bottom edge of the roofing surface the water sometimes sticks to the roofing material. It then flows around the edge and starts to move back up along the bottom of the roofing, wetting the roof structure underneath.

There are various ways to prevent this from happening. You can **use a drip edge or flashing to encourage the water to fall off.**

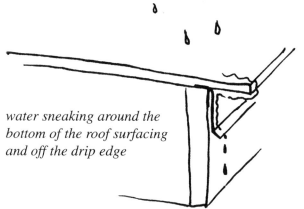

water sneaking around the bottom of the roof surfacing and off the drip edge

The steeper the roof, the more likely it is that the drips will fall right off.

If you want to, (or have to) collect drinking water or water for your garden from your roof, **choose a roofing surface that is least likely to add toxins to the water.** You may want to put the water through a filtering system.

136

Some roof surface options

thatching

Thatched roofs are made of reeds, grain or grass stalks attached to a wood frame beneath with string, or with a wire screwed into a roof sheathing. Plants that grow in water are best for thatching because they are naturally water-resistant, and are therefore more durable. Some of the Native Americans in the Northwest thatched with cedar bark.

Only the bottom ends of the pieces of thatching material are exposed. The water drips from one end to another from the top of the roof downwards until it drips off the edge of the roof. Except for the ends, most of the stalks never get wet. A thatch roof needs to be at least a 4 in 12 pitch to ensure that the water will drip from one stalk end to another. The top part of the thatch will require extra care and attention and will need to be replaced more often than the rest.

In parts of the Orient, Africa, and Latin America thatch is still commonly used, but in the western world thatching is not a widespread art anymore. There are few people in Europe who still practice it, and only a handful in America. Like any art, the best way to learn it is to watch someone who knows how to do it, and then try it yourself. I suspect that thatching will enjoy a renaissance and it will become easier to find someone to teach you how to do it.

A common fear about thatch is that it will burn. By putting a 'solid' wooden or plaster ceiling under the thatch, the fire danger will be greatly decreased because the air needed to fuel a fire will be mostly blocked off. With a 'solid' layer under the thatch, you must secure the thatch with a wire that's attached with a screw to the wood underneath.

Pros: Thatching is a wonderful roofing material. It provides insulation as well as a water-repellant surface. It is lightweight, so it requires a minimum of wood to support it. Thatch looks perfect with cob walls. And it breathes! A good thatch roof can last up to 60 years with proper maintenance. Surprisingly, thatch roofs have been used successfully by people living in very wet climates.

Cons: Hardly anybody knows how to make thatch roofs, and many of those who do want to keep the information to themselves. It is very expensive to hire someone to do it for you. It takes time. It takes a lot of materials which are sometimes hard to find. If you leave seeds in the thatching materials, birds and mice might move in and damage the thatch. It is difficult to catch the water running off the roof, so you need really wide gutters. If the gutters can't catch all water as it pours off the roof, lay down some gravel and/or plant foliage to help cut down the amount of water that will splash up and hit the walls where the water hits the ground. Thatch is flammable so it may be hard to get code approval and insurance for a thatch roof.

137

slate

Some of the old buildings in Europe were roofed with large thin pieces of stone laid as shingles. In parts of England 'man'-made slate stones are available.

Pros: Slate roofs are very charming. This roofing material will last forever (in human-time) as long as the structure and roof support hold up. The stones won't burn. The spaces between the stones allow the roof to breathe.

Cons: This kind of roof requires lots of support because it's obviously very heavy. It would be next to impossible to find someone to teach you how to build a slate roof, so you would probably have to rediscover it on your own. Finding the stones for the job could be very challenging.

ceramic tiles

If you were very ambitious, had a good supply of clay, a kiln, and lots of fuel, you could make the tiles yourself.

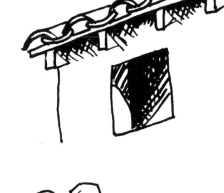

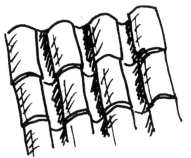

Pros: The airspaces between the tiles keep air circulating well. Ceramic tiles are non-toxic. They last a long time with very little maintenance. They will not catch on fire.

Cons: Ceramic tiles are expensive to buy. They require a lot of embodied energy to fire them. A tile roof is heavy and needs a sturdy roof structure to support it.

composite tiles

There are various commercial tiles made of different composites of plastic, polymers, cement, and fiberglass. These are expensive and have different life-spans and toxicity levels. Do some research and find out what's available in your area.

tire tiles

Tire tiles are made out of car tires (with the side walls removed) that are cut into six tiles. I'll tell you how I've seen it done: A jig saw with a razor-type blade was used to cut the side walls off, then the tire was clamped down securely and sawed into tiles with a chop saw. There was lots of yucky black smoke. Once cut, the tire tiles are laid the same way as ceramic tiles and screwed onto wooden sheathing with non-rusting screws.

Pros: Recycling old tires is next to godliness. They last a long time. They are reasonably lightweight compared to many roofing materials. They provide some insulation.

Cons: Making the tiles is a nasty, dirty, potentially dangerous job. There are a few folks who manufacture them. There is some concern about how hard it would be to put out a tire fire. They might emit various kinds of toxic gas.

sod

The dictionary defines sod as a section of grass-covered surface soil held together by matted roots. It doesn't mention anything about sod roofs. I guess these roofs are too old fashioned for the dictionary.

The basic idea is to put dirt on your roof and have plants growing up there.

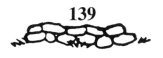

You need a water barrier between the dirt or sod and the roof structure. Traditionally, layers of bark were used. These days people sometimes use layers of plastic sheeting. It must be put on carefully to avoid piercing it. Because the plastic is protected from the sun by the sod, it will last a long time. The oldest plastic-sod roof I know of is 15-years-old and it's still working fine.

There are also fancy plastic, rubber, and bitumen products that can be used for the liner that last longer than plain old plastic. These are probably the safest membranes to use if the roof is not very steep, because you can order it in one piece. A cloth and paint roof may work under sod. (See next page.) Another product that can be used for the water barrier is called 'torchdown'. It is a rubberlike asphalt sheeting that melts onto your roof sheathing when you torch it with a flame-thrower. But it is expensive, costing about $50 a square foot, and requires a certain love of burning.

The roof has to be steep enough for the water to drain off of it, and shallow enough for the soil to stay in place until the roots grow together into a mat. A board is used at the bottom of the roof to keep the sod from falling off at the edges. You can run the liner up the board and make a drain hole through it and through the roof sheathing into the gutters. You may want to slant the roof slightly so it drains to a corner of each roof plane.

One way to make a sod roof is to cut pieces of turf out of the ground and set them up on the roof. You can make your own sod, in place, by shoveling top soil up onto the roof and throwing seeds on it. The soil can be as little as 3 inches deep.

Another way to make your own sod is to set straw bales on the waterproof membrane. Tie the bales together around the edge of the roof. Then you can cut the baling twine or wire that holds each bale together. With time, the straw will decompose and the seeds that are in the straw and windblown seeds will sprout themselves. Or you can add some manure to hurry the composting process and seed it yourself.

People sometimes fantasize about flowers and vegetables on the roof, but it usually doesn't turn out that way. Gardening on a roof is kind of awkward. The plants dry out quickly because it is such a thin layer of soil. Water is very heavy and not very insulating. It's best to sow drought-resistant plants so you don't have to water the roof to keep it alive.

Pros: A sod roof is very charming and looks right on a cob home. It helps replace the biological area that the house displaced, and is great camouflage from the air. It is easy to make and maintain. This type of roof is cheap if you use a plastic liner. It will do a great job protecting you from the heat of the summer sun, and has some insulation value in the winter too. A sod roof can have a very low pitch if you have an adequate liner.

Cons: That yucky waterproof membrane. A high quality liner can be expensive. Sod, especially wet sod, is heavy so it needs a lot of roof structure to support it. When the sod is wet, the insulative value decreases considerably. If you live in a cold climate you may want to add some extra insulation, as well as a reflective material on the ceiling to keep in your internal heat.

wood shingles or shakes

Pros: If you make them yourself, they'll cost very little money. There's a special shingle making tool you can get that makes the job easy. They are beautiful and natural and look great with cob. They can be used on an organic shaped roof.

Cons: If you buy wood shingles, they are expensive and you don't know what kind of destruction was done when the wood was harvested. They are very burnable! They need to be on a fairly steep roof so the water will drain, or the shakes will start to rot.

cloth or canvas and paint for the roof surface

This is a simple roof idea I read about and plan to try next year. It sounds so easy and cheap I thought I'd pass the idea along. I learned about this roof surface in an article written by Rev. J. D. Hooker in *Backwoods Home Magazine.*

You start at the bottom of the roof, laying a strip of cloth or canvas (the wider the better) along the roof, and tacking only at the top of the cloth. Pull the edge of the cloth around the edge of the roof. Then nail it on to the underside of the overhang with a small strip of wood. You may want to extend this cloth to line homemade wooden gutters. Paint the cloth with an exterior paint (store bought or homemade). (See the section on page 165 for recipes). When applying the first layer of paint, put plenty on so it goes through the cloth and sticks the cloth to the roof sheathing.

Lay the next piece of cloth onto the roof, overlapping the last piece by about eight inches, tacking it only along the top edge of the cloth. Wrap the sides around the edges of the roof and nail the cloth to the underside again. Paint this cloth. Continue until the roof is covered. The ridge piece is not tacked down at all, except where it wraps around the edges of the roof at the ends of the cloth.

When you get the roof covered with painted cloth, go over the whole roof with another two coats of exterior paint, letting each coat dry before applying the next.

When the roof starts looking a bit faded, you can put on a new coat of paint. If you use better quality paint it will require less maintenance. A cheap paint may need a new coat every year or two; a good quality paint may need repainting every ten years or so.

Pros: Cloth and paint roofs are cheap, if you scrounge the cloth and the paint (or make the paint). They are lightweight, durable and easy to maintain. They are easy to put on and will mold well to the shape of an organic or cone roof.

Cons: The paint or the linseed oil in homemade paint could be toxic. It would also be expensive if you bought new, good quality paint and canvas.

metal, tin

Some cautions about metal roofing: **the edges of the metal are very sharp and can cut you.** Wear gloves and don't let pieces of roofing fall off the roof onto someone or something! Watch out for the wind! It can make pieces of metal roofing into flying guillotines. Store it with weight on it and do not work with it when the wind is up.

Pros: Metal roofing is fast and easy to put up. It's strong and helps brace the rafters to each other. This gives you the option of eliminating the sheathing. Sometimes you can get off-cuts from big construction jobs for free or very cheaply. (The cut edges will be extra sharp!) Metal roofing doesn't burn, and is less toxic for water collection than asphalt roofs. You can recycle the steel when you replace the roof. Snow slides off it easily. It bends and twists a little into organic shapes.

Cons: The production of steel is hard on the earth, and metal is more expensive than asphalt shingles. It rusts quickly if it's used near the sea.

asphalt shingles

Roofing felt-paper is usually stapled onto the roof sheathing under the shingles to prevent leakage. The asphalt shingles are made of fiberglass, asphalt and other questionable 'man'-made materials.

Pros: They are easy to put on and comparatively cheap. They flex (if they're not too cold) to organic shapes fairly well.

Cons: The shingles are probably pretty toxic to produce and may let off toxic gas. They're very poisonous when they burn. Different grades of shingles are guaranteed to last for 25 to 40 years - then they make a yucky mess when they're taken off the roof, because they aren't recyclable or biodegradable.

rolled roofing and felt paper

This is basically the same stuff as the asphalt shingles in big rolls. It needs the roofing paper under it too.

Pros: It's very cheap and easy to apply. It can be used on a roof with a low pitch. (That means not very steep.)

Cons: It falls apart faster than the shingles do, but still isn't biodegradable or recyclable. Like asphalt shingles, it's toxic to produce and probably lets toxins into the air. It comes in rolls that are so heavy they are hard to get up onto the roof.

asphalt

This is a standard roofing surface on some commercial buildings. You can hire someone and their tar-heating machine to roof your house.

Pros: Hot tar and gravel are very good for flatter roofs.

Cons: It's toxic and it stinks. It burns. You'll probably have to hire someone to do it.

an old roof idea:

I came across this idea in an article by E. Crocker, about a historic hogan roof on an old Native American dwelling in the South-Western United States. It is a totally different concept than we are used to. The gently sloped roof was made up of layers of different types of clay (some absorptive and some not) laid on a log and stick framework. The strategy seemed to be to absorb the water in the absorptive clay layers, and to contain it in the roof with the non absorptive clay layers, letting it dry out between rains. I have never tried this type of roof but I think it's an idea worth exploring further.

Insulation

Insulation can be anything that creates still air spaces. Lots of roof insulation makes a house comfy. In most climates, insulating the roof is extremely important.

The temperature inside your house will always try to equalize with the outside temperature. A generous layer of roof insulation will help a lot in keeping heat inside when it's cold out, as well as keeping the hot summer sun from making your house too warm. So insulate your roof!

How much insulation?

Check with local builders to find out how much is suggested for your area. They will probably tell you something like, 'an R-value of 35'?! R-values are simply units of measurement used to describe how much insulation-value a particular material will provide. A higher number means a greater 'resistance to heat-flow'. The building supply store can tell you what the R-values are for the insulation materials they sell. You'll have to do more research to estimate the R-values of less conventional insulation materials. When in doubt, more is better.

Making a space for the insulation and its ventilation

(See page 132 for more on designing the space for the insulation.)

Where there is a moisture barrier and a temperature differential on either side of it, you will get condensation! **If you are using a roof sheathing that is a moisture barrier, regardless of which type of insulation you choose, it's a good idea to create ventilation for it.** Design the insulation space to be 2 or 3 inches deeper than the amount of insulation you plan to use. This extra space above the insulation is for air to flow through, taking any condensation with it. The top and bottom of this space need to be vented out of the roof through the cob walls. To make vents in the cob walls, bury a three-inch pipe running through the cob. Cover the pipe with screen mesh to keep critters out of the roof. The edges of the screen can be temporarily attached with a rubber band, then cobbed over along with the pipe.

These vents need to be **between each rafter** so that every insulation space is vented.

For ventilation on the tops of gable roofs, there are all sorts of fancy spinning metal vents available at building supply stores. It's easy to make your own venting, though. Leave a space at the ridge of the roof between the top pieces of sheathing and the tippy-top of the roof skeleton. This creates a vent opening along the entire top of the roof. Staple heavy-duty mesh along the opening to keep the critters out. Build a little roof (8 to 10 inch overhang) along the top of the roof over the opening, or buy the special flashing that's made to keep rain out.

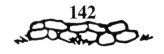

If you will be making a flat ceiling to create an attic space, make large vents from the attic space to the outside.

Mice and rats love to live in the attics or insulation spaces if they can get in. These places are warm, safe from predators, and have a wonderful supply of nesting materials. Your first line of defence is simply not to build or leave any entrances! Cob is so thick and hard the critters will have a hard time drilling through your walls. Use extra heavy duty mesh to cover all roof vent openings, and don't leave any gaps where wood meets cob or where wood meets wood.

Insulation options

Thatch The air in and between the tubes of grasses or reeds that make up the thatch are great insulation in themselves. The thatch is usually about a foot thick. In cold climates you may want to add extra insulation under the thatch.

Sod The earth, the plants, and their matted roots provide some insulation on a sod roof. It's probably a good idea to supplement it with some other type of insulation and/or reflective material, because when the sod gets wet the insulative properties decrease significantly.

Straw If you use straw, make sure it's straw and not hay. Hay is the nutritious part of the grass and is much more inviting to critters and mold than straw is. Straw bales placed close together in the insulation space have fantastic insulative properties. They are fairly heavy and require a fairly substantial roof structure of wood or steel to support their weight. They're cheap. If the straw was grown organically and if it was kept dry and mold-free, it should be nontoxic. Straw bales will smolder if exposed to enough heat, but won't really catch on fire if they can't get enough oxygen.

Loose straw has fairly good insulative properties, but one of its major drawbacks is that it burns well.

Light-clay straw One way to prevent straw from burning so well is to coat it in a clay slip. Put water, and the purest clay you have, into a big bucket or drum. Stir until the clay is suspended in the water - about the consistency of thin cream. Sand and stones will fall to the bottom. Push the goo through a screen to clean the mix and break down all the clay. This is called clay slip. Throw loose straw on a tarp and pour the clay slip on it. Use about the same proportions as salad dressing to salad. Two people can grab the corners of the tarp and toss the straw 'salad' and clay slip 'dressing' until each piece of straw is coated with the slip. Let it dry a little, then put it loosely into the insulation space. Use at least ten inches of this kind of insulation. It will provide you with about 2 or 3R of insulation per inch. Remember to leave a gap (at least 2 inches, more is better) over the straw-clay for ventilation.

Wool Combed wool has about the same 'R-value' as fiberglass. I imagine, in the not-too-distant-future, it will be easy to buy wool insulation commercially, as people become less willing to live with toxic fiberglass and chemically treated cellulose. Wool insulation batts are made in New Zealand but the U.S. requires toxic treatments before importation. Is it fumigated? If so, are you willing to live with it? Consider doing it yourself. Sheep farms often have dirty or scrap wool they're happy to get rid of. Cleaning the wool is important, to discourage critters. You'll need to brush (or card) the wool to fluff it up and make the airspaces that give it its ability to insulate. Cleaning and brushing the amount of wool you'll need is a big job. You can put the wool in loose in your insulation space, or you may choose to first stuff the wool into agricultural/burlap bags or tubes.

143

Cork Cork is bark that can be harvested without killing the cork trees. Cork comes in granular form or pressed into sheets. Untreated cork seems to be a very non-toxic insulation material. It may be hard to find and will be expensive.

Vermiculite Vermiculite is natural chips of mica processed to make them puff up. It's commonly used in gardening soils or to insulate concrete masonry. It is fairly expensive if you buy enough to insulate a whole roof. If you're not allergic to it, it may be worth the cost. Any dust is bad for your lungs, so wear a mask when you are handling it. Vermiculite doesn't burn. It will absorb moisture and make a nasty heavy mess before you realize there's a leak!

Pumice Pumice is a volcanic rock that is so full of air pockets it floats on water. I have never tried it but if you live near a pumice deposit, you might want to try using it to insulate your roof.

Wood Sawdust or wood-chips have been used for centuries for insulation. You can imagine some of the problems: fire, rot, insects, etc. Recently, some low-toxic treatments for wood chip have been developed in Europe to minimize these problems and may soon become available elsewhere.

Seaweed When I was in Australia I heard that seaweeds that have flotation bubbles were used in a lot of the old homes for ceiling insulation.

Cotton Cotton is a possible insulation material, although the production of cotton is very hard on the environment. Let us know if you figure out a good way to use recycled clothes.

Ground up newspapers (chemically treated, blown in cellulose) In Southern Oregon, most of the roofs in new subdivisions are insulated with shredded newspapers that have been treated with something (probably toxic) to reduce their flammability. This is blown into the insulation space or attic with a big machine. If you decide to use untreated shredded newspaper, be aware that it is a great fire starter.

Soy milk containers Some friends in New Mexico insulated the roof of their earth home with soy milk containers. They cut them in half, washed them, and stapled them together. These were set into the insulation space face down, towards the heated house. They report that this works as well as the fiberglass insulation (about a foot thick) in an identical house next door. You could try using whole cartons.

Cardboard My grandmother's house was built in the 1920s. It wasn't super well-insulated, but it was reasonably warm in the winter with one woodstove. In 1995, the house was smashed by a big tree in a wind storm. Luckily my grandma was OK!
I watched as the bulldozers tore down what was left of the house and I noticed that a lot of the house was made out of layers and layers of cardboard glued together. I didn't see any mouse tunnels in it, even after 60 or 70 years!! Some of the glues used in cardboard might be toxic.

Fiber glass Fiber glass insulation is made out of yucky, itchy, sharp fibers that may be stuck in your precious lungs forever! They make your skin itch and they get stuck in your eyes and nose. Some brilliant company now sells fiber glass insulation in 'poly bags' which contain the fibers, and help protect people using the stuff.

Blown-in fiber glass There are big machines that can blow fibers of spun glass. This is usually used in a situation where there is an attic space. Do you think yucky, sharp fibers flying through the air is a good idea?

Silver bubble wrap This is what it says it is. It's expensive and would require more than one layer to really insulate a roof enough. The reason I mention it here is that it is very flexible, so it works well for an organic shaped roof. It takes up very little space for the amount of insulating value it gives. It is also reflective. It could add an insulating and reflective layer to augment some other kind of insulation. It can be stapled onto the bottom of the sheathing and/or laid just above the ceiling. Pieces of this stuff lining the backs of window shades or against skylights really helps keep the heat in on cold nights. Bubble wrap is often used to wrap water pipes to keep them from freezing.

Rigid foam This is a commercially available product made out of urethane foam. It's sheets of styrofoam-like stuff. It is extremely toxic when burning and probably not that good for you to live near even when it's not burning. This stuff is expensive. One of the good things about rigid foam insulation is that you get a lot of insulation value in a small amount of space.

Anything that has trapped air spaces in it Use your imagination! What about old film canisters with their lids on? If you discover a natural or nontoxic recycled material that can be used for practical insulation, your name will go down in natural building history! Please let me know!

A reflective layer reduces the amount of insulation you'll need

Shiny stuff reflects some of the heat back. Put the shiny side towards the heat that you want to reflect. This means shiny-side down toward your room to keep heat in where it's cold, shiny-side up to reflect the sun's heat away in hot climates, or both. Put the reflective layer between the heat source and the insulation. The heat will be reflected regardless of what is on top of the reflective layer.

Silver bubble wrap and silver paper, both available at supply places, can be used for a reflective layer. One free source of reflective material is soy and rice milk containers, opened, flattened, washed, and stapled in place. In some countries, milk that doesn't need to be refrigerated comes in those kind of containers. This is a useful thing to do with a hard-to-recycle product.

Another way to reflect the summer sun is to paint the roof surface with a reflective white paint.

Ceiling

See the illustrations on pages 123 and 132.

Purposes of a ceiling

The ceiling's job is to look good and to cover the (usually) ugly insulation or reflective layer. Ceilings can also be used to hold the insulation up. In a house with an attic, the ceiling serves to separate the attic from the rooms below. If you are using something toxic for the insulation, it is important to choose a ceiling that will separate you from the toxin as much as possible.

Getting the insulation and ceiling up there

If you are building the kind of roof where the insulation sits in the roof structure, **put the insulation in place and the ceiling up (or whatever will support the insulation) before you do the sheathing.** Ideally, you will have enough time before wet weather to get this done. This way you can stand on the rafters and work with gravity, setting the insulation *down* into place on the ceiling, instead of standing on a ladder and trying to get the insulation to stay *up*. Doing things in this order will postpone the day your home is protected from rain, but will save you time, effort, and a sore neck. Use your judgement and weather forecasting skills.

If you put the roof sheathing and surfacing up first, you will get a lot of neck and arm exercise when you put the insulation in from underneath. Getting everything to stay up there requires ingenuity and patience. If you are using sheets of something for the ceiling, you can put them up one at a time, starting at the lowest side of the roof's slope. Then stuff the insulation into the space that you've created. You can use strips of wood, lath, sticks, burlap, strands of wire or chicken wire to hold the insulation up. These can be sturdy enough to provide permanent support for the insulation, or just strong enough to hold the insulation up there until you get your supporting ceiling attached.

If the ceiling is holding up the insulation, it will have to be a strong enough material to do so, and it will have to be attached securely to the bottoms of the rafters. Use your common sense.

Ceilings can be attached to the bottom of the rafters with nails or screws. Working over your head like this is challenging for your body. Do any part of the job you can on the ground, (like painting, drilling holes, and starting the screws), before putting the ceiling up - this will save your neck, arm muscles, and eyes.

You may want to cover your insulation-support with something pretty, like fabric, reeds, woven sticks, and/or plaster. If you want to plaster the ceiling, make sure that whatever you use to hold up the insulation will also support the plaster. Ceilings do get dirty, so you may want to repaint or refabric occasionally.

If you are using poles or pieces of wood (like a small beam system) to hold the ceiling up, you could leave them exposed so you can look at them, and then put the ceiling on top of them. You will need to build your rafter system to create the insulation space and to hold up the roof. This is a very sturdy way to build and it's easier to attach the ceiling *down* to the tops of the ceiling supports than *up* to the bottoms of them. If you have an abundant ecological wood source, you might want to build your roof system this way.

You can leave a ledge in the cob to help you hold the ceiling up there while you attach it to the rafters. Put the ceiling up and cob your walls up to it.

Possible ceiling materials

Materials that are strong enough to support the insulation. Use common sense.

- **boards**
- **tongue and groove boards**
- **plywood**
- **sheet-rock**
- **strips of wood**
- **wood lath**
- **strands of wire**
- **wire lath or chicken wire** (Some people don't like living with metal around them because it changes the natural magnetic field.)
- **latias:** small sticks, reeds or bamboo laid next to each other on top of poles or pieces of wood. These can have a layer of cob on top of them to isolate insulation 'dust' from the house, or they can be plastered over from the bottom if you prefer the look of plaster.

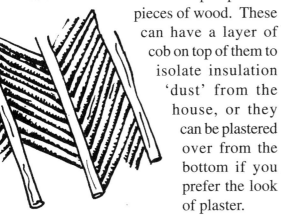

Ceiling materials that are used mostly for looks, to cover whatever is holding up the insulation

Use your imagination! Here are some ideas:

Plaster Plaster covers the insulation and helps isolate the insulation 'dust' from the room.

If you plaster the ceiling, the plaster will need something rough enough to stick to and strong enough to support it. There are various things to attach to the rafters to hold the plaster up there, for example: wire lath, sticks nailed on top of the rafters, woven sticks, thin wood sheathing with holes drilled in it, and pegboard. (See the picture on page 101.) These can be nailed, tied, or stapled to the rafters.

Certain plaster ingredients have more air in them than others. Vermiculite, pumice, and straw make small air pockets in the plaster which add a small amount of insulative value. These can be put into the ceiling plaster if you need a little extra insulation up there. Make various tests to help you decide which combination of ingredients will stick best to the ceiling.

Homemade sheet-rock

This is a concept that is still in the experimental stages. The idea is basically plaster with a fabric core. You could play around with cloth (old sheets work great) dipped in clay slip or wet gypsum plaster. Depending on what you've used for insulation, this lightweight 'sheet-rock' may be able to stick right to it. Adding a little glue to the plaster or slip will help. Or you could attach it with wood strips nailed through the fabric into the rafters. You can also put the fabric up first and paint it with the clay slip or homemade paint. Experiment until you find something that you like that works.

Another homemade sheet-rock idea is to use layers of magazine pages or newspaper, instead of using fabric. These could be dipped in clay slip and stuck onto each other. I've only seen this in an experiment, but I think it might be worth trying for a ceiling or an interior wall.

Thin plywood or veneers

These could be left exposed, plastered, painted, or covered with wallpaper. You could play around with making your own homemade wallpaper or papier mâché.

Fabric

You could use fabric stapled up to the rafters to cover the insulation. This provides a very lightweight, beautiful, organic shaped ceiling. It may need to be replaced, washed, or painted over in a few years when it gets grubby.

You might want to keep an eye on the roof system to make sure it's functioning properly. (No leaks, is there enough insulation, are those mice having fun?) So it may be a good idea to use a ceiling that is easy to take down and look under, at least temporarily.

Woven mats or reeds

You can buy long rolls of reeds or bamboo woven with metal wire. These are ordinarily used as window shades or fences. These make a very pretty, natural-looking ceiling and can be attached with strips of wood or staples.

Bamboo, twigs, or reeds

These can be nailed or tied to the rafters, or woven together first and then attached to the rafters. These can be plastered over.

Wood

The pieces of wood for a ceiling only need to span from one rafter to another, so they can be short. This is a wonderful way to use up the little scrap pieces of wood from mills or construction sites. You can run a long strip of wood where all the little pieces of wood meet if you want to cover the seam.

PLASTER (RENDER)

Purposes of plaster

1) On the exterior walls, plaster protects the cob from erosion by wind, rain, sun, and snow. Exterior plastering on cob is only necessary if you live where there is wind-driven rain, or snow buildup beside the house.

2) Plaster gives the walls a smooth finish.

3) Plastering reduces dust.

4) Plastering is a great excuse to have a party! I wonder if that's where the saying "getting plastered" comes from? When you plaster, you caress every inch of your lovely new home or re-caress every inch of your lovely old home.

The line between cob, plaster, alís, paint, and washes is nonexistent. One graduates into the other as the amounts of water and particle sizes vary. When you see the word alís, it refers to a fine plaster or a thick paint. With natural buildings it is vital to use a plaster that breathes and moves. These plasters will absorb and release moisture, allowing the walls to dry out quickly. Vapor must be allowed to escape from interior spaces and the building materials themselves, where they could cause damage.

A general rule of natural building is: like sticks to like. An earth-based plaster is perfect for a cob house. Lime has also been used around the world to make a breathing plaster or a paint to protect and beautify earth walls.

Cement-based plasters or stuccoes are usually destructive to earth buildings. The hard cement coatings almost always crack because their rigidity makes them brittle. The cracks let moisture in, wetting and often eroding away the earthen structure under the crack. The cement coatings breathe less and slow down the drying process of the cob beneath them. Many old earthen buildings that were 'restored' with cement based coatings are now having to be redone. All the cement has to be removed, the damage repaired, and the building replastered with more compatible, vapor permeable plasters.

Mixing plaster (render)

Plaster can be mixed on a tarp just like you mix cob. If you want to make really big mixes, you can make them in a pit dug in the ground. (See page 75 for straw bale pit ideas.) Smaller batches can be made in a wheelbarrow or in a bucket or tub. Cement mixers also work well for mixing earth plaster, if you can stand the noise!

tools you can use for plastering

- putty knife
- hawk (see page 158)

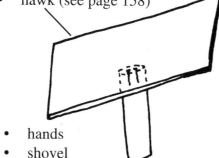

- hands
- shovel
- sifting screen, fine sieve
- tarps
- tubs or buckets
- wheelbarrow
- straw bale pit
- plastic yogurt or cottage cheese containers
- smooth polishing stones

- sponge or sheepskin
- big paint brush
- hose with a spray nozzle
- squirt bottles and spray bottles
- whisk broom
- trowels

you can use a little trowel for mixing the plaster, for applying the plaster and /or for smoothing it

A swimming pool trowel works well for applying the plaster on large areas and is great for putting down a cob floor.

Basic earth plaster

Now that you've become familiar with your soil and with mixing cob, making an earth plaster will be easy. **A basic plaster is simply a finely sifted, wetter version of cob.** It is made of the same ingredients in the same proportions as cob: sand, clay, and straw (or some other fiber). Use the purest clay you have for the plaster. Put everything through a fine screen and add more water. That's it! This plaster mix can be used for a lovely, durable finish on cob walls.

Sifting the ingredients

If you have large particles, pebbles, or a lot of organic debris in your soil and/or sand, you will need to sift it. A sifting screen can be made by stapling a screen onto a wooden frame, or by nailing strips of wood over a screen onto its frame. You may want to make more than one screen, using different sizes of mesh. If you are making more than one layer of plaster, the under layers can have larger sized particles (1/8 inch mesh). Only the outer coat of plaster will be seen. If you want a fine finish, you'll want to sift all ingredients through an ordinary window screen or an even finer screen.

You can lay the screen over a wheelbarrow or bucket and sift into that, or lean the screen up against a straw bale or wall and sift onto a tarp. The soil can be sifted dry after crushing up the lumps with your shoes or with a tamper if necessary. Or you can mix the ingredients with water, let it soak a day or two, and then push the goo through the screen by rubbing it with a smooth rock or your hands. Dust is kept to a minimum if you do the sifting when the soil is wet.

Shredding the straw

Straw for the plaster also needs to be sifted or shredded. Straw dust can irritate lungs and eyes. Wear a mask and eye protection when breaking it up. It's much easier to shred when it's very dry (not damp). You might want to leave the straw in the sun during the hottest part of the day and then shred it.

The straw can be shredded by placing a 1/8 inch mesh screen on top of a 1/4 inch mesh screen, laying them over a wheelbarrow or bucket, and **'grating' handfuls of straw through the mesh.** The finer you want the straw, the finer mesh you'll need. Straw can also be shredded **with a lawn mower** by running over and over it. Or you can use a **weed eater** in a metal drum to break up the straw. Other methods include: old blender, chain saw, electric knife, drill with wire brush, scissors, and chopping with sharp knife on a wooden board.

For the initial plaster coats (sometimes called rough coats), the ones that will be covered later, you can use the straw lying on the floor of the cobbing area, that has already been **shredded by feet** and tarps being dragged over it. I purposely throw extra straw down to be shredded for the plaster mixes. You can use a roughly shredded straw in the final coat of plaster if you like the look of the textured finish and the bright, straw pieces reflecting light. Make a test batch so you can see whether you like it.

For a very fine outer coat of plaster, **you can replace the straw with fresh or finely-grated dry manure.** The digestive systems of grass eating animals have already done the job of shredding the fibers for you. The dry manure will need to be 'grated' through a sieve to break it up. Fresh manure can be added in as is.

Plaster additions

Purposes of plaster additions

Additions to the basic earth plaster recipes are not necessary, the basic fine cob works well. Additions serve one or more of the following purposes:

- Making the plaster easier to apply.
- Making it more water resistant, discouraging erosion.
- Increasing its ability to stick together and to the wall.
- Adding strength to the plaster.
- Enhancing the plaster's beauty.
- Minimizing cracking.
- Adding magic things adds magic.

Make lots of different test batches.

Experiment with various additions. You can add one or more to your basic cob plaster mix. Put the different mixes on the walls. (See page 157 for more on applying the plasters.) Thin layers of plaster are best - at most 3/8 of an inch thick. You can label them by writing the ingredients right into the plaster. These test spots can easily be plastered over later. Let them dry and have a look at them. Which ones were easiest to apply? Which ones look the best to you? Which ones cracked? **As with cob, if your plaster cracks, add more fiber or sand.**

Possible additions

plant fluff: is a strengthening fiber, sort of like felting. It can be used in addition to or in place of the straw, and discourages erosion, adds magic, and minimizes cracking.

Fluff can be gathered from any type of plant that has seed fluff. It's very easy to collect a lot of this quickly from cattails.

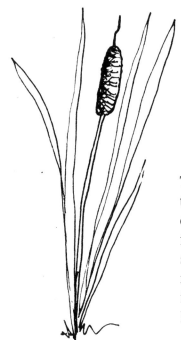

This is one of my favorite things to add to the basic earth plaster. The tiny fibers add a lot of strength to the plaster mix. Another version of fluff can be found in the lint screen of your dryer.

hair: strengthens sort of like felting, discourages erosion, minimizes cracking, adds magic - especially if it is cut from human or animal friends. It can be used in addition to or in place of straw.

Hair is a very strong fiber. Animal hair or people hair cut into short (1/2 inch or less) pieces works well. It is hard to get it to mix into the plaster and it gives the finish a slightly hairy look. Try a little just for fun. This will make life interesting for the archaeologists of the future. Hair is an excellent fiber to add to lime plaster, because it is so resistant to lime's caustic effects.

manure from grass eating animals: strengthens, discourages erosion, makes it stick, minimizes cracking, increases water resistance. Manure definitely adds something magic.

This is an excellent addition to earth plaster. Some plasters are more than half manure. The manure makes the plaster extra strong and easy to spread. The fresher the manure, the better it is for the plaster. This must be balanced with your ability to handle the stink. It's not as bad as you imagine. The smell will completely disappear as the plaster dries. If you are using dry manure, it is hard to break it down. It's already turned into a hard plaster. You may have to grate it through a 1/2 inch mesh screen before you stir it into your plaster mix.

I think cow manure is better than horse manure, but horse will do if that's all you have. Llama or alpaca manure is less stinky than cow and easy to collect because these animals have one special toilet area. I've heard that pig shit is a favorite in Switzerland. Deer or kangaroo poo are probably good plaster additions too.

A friend told me a story he heard from a Namibian man about how his tribe plasters their earthen domes. They collect fresh cow manure, roll it into balls the size of one's head and leave these balls of manure in the sun to 'cure' for exactly seven days. Then they are broken open (Peew!!!) and used to plaster the dome. Supposedly this is a very waterproof plaster.

> **An African recipe for wall plaster**
> 1 part manure
> 1 part clay - you may choose to use a pretty colored clay
> Add water until the plaster is the consistency you want.

flour paste: (Thank you Carol Crews for passing on this great idea!) Adding flour paste makes the **interior** plaster a lot sturdier and less dusty. It makes it sticky and easy to apply. It can be used in the plaster and/or in natural paint or alís.

Mix up a flour paste. In a jar, shake one part cold water and one part flour, until it's well mixed with no lumps. Add this to boiling water, heating and stirring until it thickens.

You can use cheap white flour or whatever you have. Do some test batches using various ratios of paste to plaster. **Start with one part paste and five parts basic earth plaster.** Experiment from there adding more or less paste. I have only used flour on the interior plaster. I'm not sure how it will hold up outside.

Once I was doing a demonstration at a workshop and we ran out of flour. We did have some leftover cooked oatmeal from breakfast so we blended it up and added it to the plaster. It worked beautifully.

cactus juice: makes it stickier, increases water resistance.

This is commonly used where cactus grows. Every region has some plant that has mucilaginous juice that may serve the same purpose as the cactus juice. Check which varieties are used in your local region, how they are prepared, and what proportion of juice is added to the plaster. The cactus is usually peeled, de-spined, cut into pieces, boiled, and then strained. The goo is added to the plaster in place of some of the water. Experiment with amounts. **Start with 1/2 a big bucket to a wheelbarrow of plaster.**

ground psyllium seed husk: makes the plaster a joy to put on, holds the plaster together well, increases water resistance, and discourages erosion.

This can be bought in small quantities at health food stores. It is ordinarily used to clean out the colon. **You'll only need a little bit. Start by adding a quarter of a cup to a 5 gallon bucket of plaster.** Adding psyllium seed husk to the mix is kind of like adding rubber. If you add too much, the plaster will 'glom' into a ball and only stick to itself. Psyllium is a wonderful addition to earth floors.

Elmers glue or other manufactured polymers: adds water resistance, makes it stickier.

This is a great addition for a plaster that will have tiles or other heavy decorations embedded into it. **Try 1/2 cup (or more) of glue per big five gallon bucket.**
Some folks have been doing experiments using mixtures of different types of glues and natural ingredients. There's still some experimenting to do before we'll know which kinds to add to the plasters, how much, how well they work, and whether they breathe enough to keep a cob wall healthy. They seem very tough and water resistant.

coloring: for looks.

You may want to collect special colored clays for the last layer of plaster or for a paint. **A colored final layer of plaster can take the place of paint.** You can use potter's clay as a colorant for the outer layer of plaster.

If you are adding colored dry pigments, mix them well with water before you add them. (See the section on paint, page 163-167 for more coloring ideas.)

Termite mound dirt: If you live where the termites build dirt houses, you can find an abandoned mound, crush it up, and use it as part of your plaster. I met a few people who had tried this in Australia. They reported that it made a very weather resistant plaster and highly recommended it.

papier mâché: ground-up paper, adds a small amount of insulation, helps the plaster stick together, adds tiny fibers that increase the strength and cut down the erosion.

Make some test batches with various amounts of ground-up paper. In one of our experiments, we used only ground up telephone book yellow pages mixed with water for an exterior plaster, and were amazed at how well it held up. Adding some clay and/or flour paste would make it even better.

mica or glitter: looks pretty, catches the light.

Mica or glitter added to the final layer of plaster makes beautiful shiny bits that glint in the light. These ingredients can be sprinkled onto the plaster after it's on the wall and gently rubbed into the surface. Mica is a naturally shiny, glass-like scaly stone available in the southwestern United States, and possibly elsewhere. It is like platelets of sand and may increase the strength of the plaster.

vermiculite, pumice, or extra straw: It may be possible to add insulative value to the plaster by the addition of materials that create pockets of air in the plaster. This concept needs testing. If you want to experiment, start by adding 1 part to 10 parts of plaster. See how much you can add before the plaster mix gets too crumbly. You may want to use a little extra of the sticky additions mentioned above to help hold the plaster together so it will hold more insulation.

aging plaster: Like cob, all these plasters benefit from sitting around for some time. They seem to become easier to work with and more sticky. Mix up a wet batch and let it sit for a few days or more. Yum yum!!

Applying the plaster or render

When to plaster
Interior plaster usually goes on after the roof and ceiling are completed, and before the floor is made. (Unless, of course, you are replastering long after the floor is done. In which case you'll want to protect the floor, furniture, etc., from plaster drips.) The exterior plastering can be done after you've moved into the house or you can leave the walls as they are.

Don't plaster when it's below freezing, or when there's a chance that the plaster will freeze before it's dry. If the sun is shining directly on the wall you will be plastering, set up a shade for it. It's best not to plaster when there's a dry wind blowing. **Ideally, the plaster will dry slowly.** This is especially important if you've used lime in your plaster. (See page 162 for more about lime plasters.)

How many coats of plaster?
Three is a typical number of coats on earthen structures. I've never done more than two on a cob building. Maybe it is because I am very careful as I build the walls so there isn't much filling in or adjusting to do.

Some folks like the look of unplastered cob. It can be whitewashed or a layer of alís can be applied directly to the rough wall, if you want a very textured surface. For a smoother wall, chop the loose straw ends off before whitewashing.

If you want to leave the exterior of the cob walls as they are, without plastering, observe the walls for erosion during the rainy season. If you can detect any wear, you may want to plaster or paint with lime wash to protect the cob, or build a porch or verandah on to the weathered side of the house.

You can add new layers of plaster whenever the urge grabs you. Remember to dampen the surface of the walls before applying a new coat of plaster. If you add plaster to the inside walls too often, your house will fill up with plaster. If this happens, you have a bad case of plasteritis. Please write a ballad about it and sing it to the next Natural Building Conference in your area.

Preparing the walls for plaster
Some natural builders say to let the cob walls dry well before plastering them. The theory is that the walls might shrink and crack the plaster. Also the plastering slows down the drying rate of the walls. Some old books say to plaster the interior, move in, and plaster the exterior next spring.

Cob walls take time to dry out completely. If you just can't wait, it's probably OK to plaster onto partially dry walls. I have done it and had no problems with cracking. You can always plaster over it again if it cracks.

The more similar the moisture content of the wall and the moisture content of the plaster, the better the plaster will stick. Wet the surface of the walls well. You can spray them every 5 minutes or so while you are getting stuff organized to plaster. This is your chance to make last minute changes to the shape of the surface of the walls. Use a shovel and/or machete to carve the big bumps off the walls. Layers of plaster can be used to fill in minor dips and holes. Any gaps where the cob has shrunk away from the wood frames can be filled with plaster.

You can leave the shaggy straw that's sticking out of the cob wall. Most of it will get folded over during the plastering and will help to hold the plaster onto the walls. If some stray straw sticks through the plaster, cut it off with scissors.

To carry the plaster from where it's mixed to where it goes on the wall you can: use a wheelbarrow, drag it on the tarp, carry it in a plastic dishwashing tub, or on a plaster hawk. A hawk can be made out of a dowel screwed to a small piece of board or plywood.

Methods of application

- by hand
- trowel
- piece of a plastic yogurt container (this is also a burnishing or polishing tool)
- sponge or sheepskin
- big paint brush (for a thin outer plaster, alís or whitewash)

Try putting the plaster on with all these methods. You will probably prefer one way and stick to that. **Keep your tools and hands wet so the plaster won't stick to them instead of the wall.** The general idea is to apply **a thin layer (3/8 of an inch or less),** in upward sweeps.

Start on a back wall, under the cupboards, or somewhere out of view so your first attempt won't be so visible. Practice will improve your abilities quickly. Keep at it! Your first layer of plaster will be covered so don't be too perfectionist about it. By the time you get to the outer layer you'll be much better at it.

Scoop the plaster onto the hawk, lean the hawk on the wall, and scrape a blob of plaster off the hawk onto the wall with your trowel.

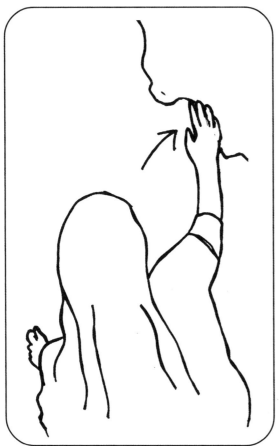

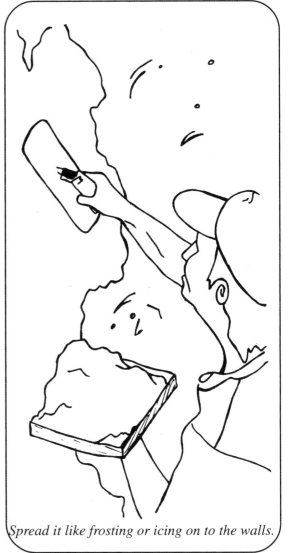

Spread it like frosting or icing on to the walls.

Where to start plastering

Some folks start plastering at the bottom of the walls and others at the top. I'm one of the start-at-the-top kind of people, because I don't like slopping on my finished work. The logic of starting at the bottom is that as you add sections of plaster to the walls they overlap each other like shingles, encouraging the water to run off the wall instead of into any slight seams that might form where the different applications of plaster join.

How thick do I make each layer of plaster?

The first layer of plaster needs to be thick enough to fill in the dips and can be quite thin over the high points. If the dips are deeper than 1/2 an inch, you may need to fill them with two or more 1/2 inch layers of plaster.

A general rule for plaster thickness is: the smaller the particles in the plaster ingredients, the thinner the layer. The thinner the plaster, the less it will crack. Do some small tests of different thicknesses in out-of-the-way places.

Don't work the plaster too much while it's wet. This brings the adhesive materials in the plaster to the surface, away from the wall where they are needed. If you plan to add another layer, don't worry about how pretty it looks. In fact, leaving it rough gives the next layer more to hold onto. If you want to level off the surface, scrape the edge of a board across the partially dry plaster.

Make sure you **wet the last coat of plaster (or the wall) before you add the next coat.** Put the last layer of plaster on with care because this is the coat you will see. When you've finished plastering for the day, try to leave the edges of the latest plastering section somewhere where they are not too obvious. There may be a slight seam where one day's work ends, and the next day's begins.

You can make relief sculpture on the walls with the plaster. Try some sculpting as you plaster. You can always chop it off later if you decide you don't like it. Sculpting around door and window openings can be very beautiful.

A whisk broom (straw hand brush) is a handy tool for plastering. You can use it

to whisk off the loose stuff on the cob or the last layer of the plaster. Keep it in a bucket of water near you, to splash water onto the dry wall. After you've put the fresh plaster on, brush the surface with the broom to smooth the plaster and to give it a lovely texture. A long handled broom can be used to reach the parts of the wall that are over your head.

Embedding decoration into the plaster

Plastering is a work of art. You can set pretty tiles, stones, sticks and bits 'n pieces into the plaster. Some people add a little Elmers glue to the plaster behind the decorations and tiles.

On the outside surface, you may want to embed small stones, seashells or pieces of mosaic tile along the bottom 18 to 30 inches of the wall. This will help protect the walls from water splashing off the roof and from the worst of the windblown rain. This can be a wonderful opportunity for the artist in you and can add an elegant as well as practical touch to your home.

You can make beautiful textures inside and out, by printing or cutting into the wet plasters. Experiment! You can always smooth it back over if you don't like the results.

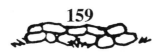

Polishing the plaster

To erase unwanted bumps and lumps, go over the plaster with a damp sponge or rag when it's partially dry and 'sand' off the raised areas. (Burlap works well.) To flatten a large area, take a piece of 2x4 or a wooden trowel and rub it in a circular motion over the plastered wall.

The final coat of plaster can be burnished (polished) to give it a super smooth look and to strengthen and compress it. Polishing aligns the clay molecules and makes it a lot more durable. This can be done instead of adding an alís or the alís can be polished. If you like the look of really smooth white walls, polish your plaster and then whitewash it.

Polish the plaster when it is partly dry. You will soon learn how much to let it dry for optimum polishing. Keep an eye on the drying. You can burnish (polish) the plaster fairly soon after you apply it.

If it gets too dry before you get to it, you can re-wet it gently with a fine spray and then polish. This is where the pieces of plastic yogurt or cottage cheese containers come in. Cut 4 inch round or oval pieces out of the sides and lids of the containers.

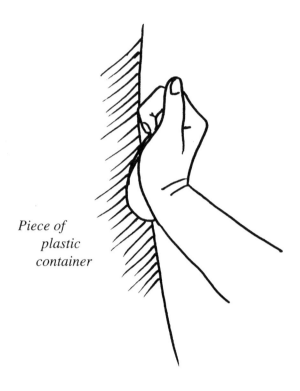

Piece of plastic container

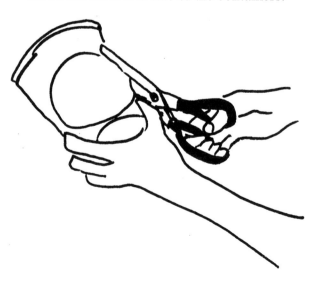

You'll need quite a few because they do wear out. These can be used to apply the plaster and to polish it.

It is especially important to polish areas that will get lots of wear, like around light fixtures, in the kitchen, and the cob furniture.

A smooth stone makes another good polishing tool. This leaves a lovely textured look on the walls.

Replastering Later

When you are replastering after a few years of living in your home, you'll find that it is easy to patch worn areas. Fresh earthen plasters stick well to old plaster. Wet the area where you are replastering. Use drop cloths to protect your counter tops and floors from splatters.

Plastering on walls other than cob

Plastering agricultural bags

Plaster agricultural bags as soon as possible because they will break down quickly in the sunlight. You'll be surprised how well the plaster sticks to the bags. For the tricky areas, like the underside of where a bag overhangs, plaster where you can, then let it dry. Your next application of plaster will stick to the dry plaster and form a bridge to the next area of dry plaster.

On brick or stone walls

Natural plasters will work well on any natural surface.

On straw bale walls

You can use these plasters on straw bales too. Spray the walls lightly with water before you plaster them. When an earth plaster is the right consistency, it will stick beautifully to the straw, especially when it is put on by hand. You won't need any chicken wire or lath. Plaster right onto the straw. Use two or three layers of plaster on the straw and finish it just like you would a plaster on a cob home. The earthen plasters breathe well, which keeps the straw from rotting.

Because lime is caustic and breaks down organic matter, it's probably best not to use it directly on the straw. If you want to use a lime plaster, you could put a layer of earth plaster on first, then try the lime. Or use an earth plaster and a lime paint.

Thick layers of plaster on a straw bale wall give it structural support and added thermal mass.

Other Plasters

Gypsum plasters

I have very little experience with gypsum. I did use it once at a workshop, but after breathing the fumes as I put it on, I felt sick and headachy. Because of that and because I'm so happy with earth plasters, I haven't bothered to learn anything else about gypsum. I'll tell you the little that I know. Gypsum is one of the main ingredients in plaster of paris and some people are allergic to it. The Gypsum we used came out of a bag at the building supply store. We followed the instructions and mixed it with water and sand. It set up slowly and can be mixed with an earth plaster. The instructions said for indoor use only.

Builders' lime plaster
(not gardening lime)

Caution: lime burns skin and eyes! Wear glasses and gloves.

Lime mixed with sand and water was the most common plaster used on traditional European cob buildings and as a mortar for stone buildings. It was also used in the plasters and mortars of the ancient Anasazi and other Native Americans.

In the United States, lime has gone out of fashion, so people there can't go to the local lime kiln to get lime or to the building supply store to get lime putty. People in the U.S. usually use dry powdered lime. This is an inferior product and loses quality when it sits around in the dry state. If that's all you have, you may want to use an earthen plaster and just use the powdered lime putty to make paint.

Making a lime putty

If all you can get is the powdered lime, mix it in water to a sour cream consistency to make a lime putty. Add the lime to the water slowly and stir it well. Let it soak for a few weeks before using. This putty can be mixed into plasters or mortar. While it's soaking an important chemical reaction occurs. Once it is mixed with water, the quality shouldn't deteriorate and the lime can be stored wet indefinitely.

Recipe for a lime plaster

1 part lime putty (See above for how to make this putty)
3 parts sand of different particle sizes
1/3 to 1/2 part fiber (cow hair was traditional in Britain)
Remember to protect your skin and eyes! Mix the putty and the sand well (you can use a cement mixer), **cover it well and let it sit for two weeks. Then add the fiber to the amount of plaster you can use each day.** Lime is caustic and corrosive to organic matter. Hair is an excellent fiber to use with lime because it is very hard to break down. Chopped up polypropylene baling twine works well too.

Lime plaster takes more skill to put on than earth plasters. Practice in some out of the way spots on the wall. Wet the walls well before plastering with lime. Apply 2 or 3 thin layers, always wetting the previous layer before adding a new one. The proper way to put the first coat on and get it to adhere well to the wall is to flick it onto the wall with a backhand stroke off a little shovel (like the kind for cleaning out ashes from the stove). When it is partly hard, go over it with a piece of wood in a circular motion to even out the surface. I tried this and never got the hang of it and I didn't like the caustic splatters, so I just put it on with a trowel and that worked OK. Make the layers as thin as possible. Do not fuss with it too much once it's on the wall.

Let it dry until it is firm to the touch but not all the way dry before you add the next layer. You can rub the plaster with a piece of wood between each coat to even it out. Scratch the previous coat when it's still damp to make it rough so the next layer will stick better.

Lime is like concrete, it needs to cure slowly. In English reference books about lime plasters, they suggest hanging wet burlap sheets a few inches from the freshly plastered walls for a week or more after plastering. Shade the fresh plaster from the sun. Do not plaster on dry, windy days.

A lime plaster makes a hard looking surface similar to a cement stucco.

About lime additions to earth plasters

Lime added to an earthen plaster makes it harder. It also lightens the color. Adding lime makes the plaster caustic, so remember to protect your skin and eyes. If you are adding lime putty to your earthen plaster, make test batches with different proportions of lime putty. Put them on the building, let them dry and see which mixes work well. Look for cracks. If you do add lime to your earth plaster, let it dry as slowly as you can. Wet the walls well before you put the plaster on. High humidity days are good for slowing down the drying process.

Alís and Paint

Alís (pronounced "a-lease" with the accent on lease) is a term referring to a thin plaster or thick paint.

The ingredients in cob, earthen plaster, alís, natural paints and washes are much the same. One graduates into the other as the amounts of water and particle sizes vary.

Do not use paints that seal and prevent breathing on cob walls! (Or on any wall made of natural materials.) The acrylics and emulsions in most modern paints will block the vital breath of a natural home! This will cause destructive moisture to be trapped in the walls.

If you paint the walls with a light color, the house will feel bigger and the light walls will reflect the sunshine and brighten up the indoors. This is always a hard decision for me because I also love the warm soft look of the earth plaster or alís.

basic lime paint or lime wash

Wear protective clothing and glasses. Because lime is so alkaline, it disinfects and discourages critters.

Mix water with lime putty until it is the **consistency of skim milk.** That is the recipe! (See page 162 in this chapter for how to make a lime putty out of powdered lime.)

This will keep very well, so make enough to cover your walls and save some for touch-ups or future coats.

Wet the wall well before painting it. Use water that's sitting on top of your whitewash if it has settled. Then add a little more water to the wash to replace what you've sprayed on the walls. Just **before you paint** with it, **stir it well and put it through a fine sieve.**

You can also use a tiny bit of lime (2 or 3%) in the water you use to wet the walls. Apply the lime paint or wash with a big coarse paint brush. The paint will look too thin, like you are painting water onto the wall when you first put it on. That's OK. It will become more opaque as it dries.

Lots (at least three, more is better) of thin layers is hardier than fewer thicker ones. Let each coat dry at least a day before adding another coat. Wet the previous layer with a fine mist spray bottle before you put the next one on.

A lime paint is a hard white color. You might want to add some pigment to soften the color. **If you are adding pigment to lime, it needs to be compatible with the intense alkalinity of the lime.** I spilled some blackberries in the dirt and not wanting to waste all the effort that had gone into picking them, I tried an experiment adding the blackberry juice to a lime wash. Well, the results were very exciting! The luscious purple red quickly turned into a turquoise, then slowly changed to green, then to a split-pea soup color and finally, after about a week it settled down to a natural light brown color. It softens the hard white beautifully and seems to be a fairly permanent color. It's kept that same shade for a year. (See page 166 for more ideas on adding color.)

After a few years, if the walls start looking a little dingy, it's easy to spruce them up by adding another coat of lime paint.

basic natural paint or alís recipe

- 2 parts fine sand
- 2 parts fine clay
- 1 part flour paste (white flour is finer)
- Finely sieved straw or manure
- water

Because of the flour, this mix will only keep for a few days. If you want to extend its life a little, keep it as cool as possible. You can mix up all the ingredients except the flour first and save some of it. Add the flour paste to it when you need to touch up the paint later on.

The sand, clay and straw or manure can be sifted through a very fine sieve for a fine paint. You can sift through one of those special ultra-fine screens sold at kitchen supply stores to keep frying bacon from splattering.

Play around with the proportions because there's a lot of variety in the properties of different clays. Try adding more flour paste to some of the test batches.

Add enough water to make the mix the consistency of a thin smoothie or milkshake. Experiment with different amounts of water until you find what you like to work with.

For a light colored paint, you may be able to find some pale colored clays in your area. If not, you can use kaolin clay. This can be bought at a pottery supply store quite cheaply in a dry powdered form. There are other light colored pottery clays. See what's available locally.

(Read the section in this chapter about adding color, page 166.)

Adding mica or glitter to your paint or alís makes lovely little sparkles on your walls. If you have a good supply of mica, you can replace up to 3/4 of the sand in the recipe with the mica. If you only have a little bit, dust it onto the wet paint or alís and gently push it in so it sticks.

Putting the alís on the wall

Let the plaster dry, then re-wet the walls. Start painting at the top of the wall so you don't drip on your finished work. The alís can be applied to the walls with a paint brush, a piece of sheepskin, plastic yogurt container, a sponge, or with your hands. Apply it with as little rubbing as possible. Let it dry until it's leather hard, then polish. (See page 160 in this chapter for more on polishing.)

Possible additions to make a more washable surface for lime or earth paints:

oil: Raw linseed or vegetable oil can be added to the outer coat of plaster, or to the paint to make it more waterproof. A little bit goes a long way. Too much oil might make too much of a moisture barrier, causing condensation to form in the wall. Try a tablespoon to a bucket of plaster.

You could try adding scented oil so you could have the pleasure of smelling it as you plaster. This could be especially helpful when applying a manure plaster or a highly fermented plaster. Just like the stinky additions, the sweet smells will fade quickly as the plaster dries.

tallow: In the old days, animal fat was added to lime washes to give a more waterproof finish. I haven't tried this but I imagine that the paint would have to be heated up to get the tallow to melt and blend in.

glue: Adding a little white glue will give you a tougher surface. A little goes a long way. Try 1/4 a cup to a bucket of paint or alís.

165

eggs: You could experiment with adding eggs to the paint. I tried painting beaten eggs over the paint after it had dried and that made the paint a lot less smudgy when quickly wiped with a wet cloth.

milk: Milk was a common ingredient in old fashioned paint. You may want to try using milk instead of the water in the recipe to make a more water resistant surface. The milk paint will not keep well.

adobe protector

Carol Crews built an adobe dome in New Mexico. She painted the top of the dome with a commercial product called 'adobe protector'. Its manufacturer says the ingredients are secret. Carol says it works well, making the water bead up and run off but still letting the building breathe. This product needs to be painted on again every year and is expensive. "Adobe protector" is carried by El Rey Stucco Products in the United States. Call 1-800-621-1801 for more information.

What about mixing earth alís with lime paint?

Mixing clays and lime seems to be a little tricky. In my experiments, some turned out dusty, others were excellent, and I'm not sure why. You might want to make up some test batches and see how it works with your clays.

A tough natural paint recipe: Not for use on cob!

This recipe is for use on cloth roofs (see page 140) or for painting wood. It is too waterproof to use on cob. Stir carefully: 1 part of builders' lime into 8 parts milk, then stir in 2 parts boiled linseed oil. Strain it through a sieve and use within a few days.

Some ideas on adding color

You can put the color in the last layer of plaster, or use a colored alís or paint.

If you are adding pigment to lime, it needs to be compatible with the intense alkalinity of the lime.

If you want exactly the same shade on the walls, make up a big enough batch of paint to cover them all. It's hard to make another batch of exactly the same color.

Be adventurous! Most of us are used to walls that are all the same shade, but varying the shades and/or the colors adds a more natural look. Sprinkling a dusting of dry powdered pigment on the wet paint and brushing it gives an unusual and natural rocklike finish. Play around, see what you can come up with. You can always change it. It's easy to paint over if you decide you can't live with your last creation.

To add color to your mix, you can use **pretty colored clays** as part of the clay proportion. It is amazing all the different colored clays that Mother Nature makes. Keep your eyes open to see what you can find. I have found that adding clay to a lime base paint makes the paint dusty. It comes off a little on your fingers when you rub it after it's dried, so it might be best to use some other colorant for the lime paint. Or maybe some types of clay work and others don't?

You can buy **natural powdered pigments** to add to your paints and alíses. These are highly concentrated and you'll only need a tiny bit to add a lot of color! Add a little bit at a time, until you get the color you want. Mix the pigment with water by shaking them together in a jar before adding it to the paint. Some natural pigments come already mixed into a paste.

Chalk-line chalk is a cheap, available, very colorful pigment that you might want to play with. It's usually bright blue, in powder form, and is available at hardware stores.

I haven't tried it, but I've heard you can use **commercial clothing dyes** although these may be toxic and dye production definitely pollutes our precious Earth. The amount you would need to color your paint would be minimal though. There must be **natural cloth dyes** available. Or you could get a book on natural dyes and experiment adding these to paints.

Laundry blueing was common in the old days for making white clothes less yellow. It makes a beautiful blue coloring for paints.

FINISHING TOUCHES

- Finish the drainage.
- Put up the gutters and down-pipes and make sure the water has somewhere to go where it can be useful and won't do any damage.
- Hang the door(s) and openable windows.
- Do the plumbing, electric, and propane. Install the light fixtures and sink.
- Put in your cooking and heating systems if you haven't already, and get firewood if necessary.
- Build in the cupboards, and finish up any work that needs to be done on the shelves.

Now you can move in!! Pat yourself on the back. Savor the pleasure!

Trails and roads

Fix up the road and the trails if you need to before the winter.

Roofs and floors for outdoor sheds

When you are through with the main house, make patios and add the roofs to the porches and sheds that are connected to the house. Put in their floors. Even a truckload of gravel or bark will look good and help keep the outdoor floor areas dry. If you want to make an outdoor cob wall, it must be protected from the rain. (See the illustration for a clever little roof that will do this job.)

Around the house

Putting bark, gravel, or plants along the side of the house will stop some of the splash caused by drips off the roof, which can erode the bottoms of the walls over time. Like any type of home, it's not a good idea to let the sprinkler run onto the walls.

A clay type soil will shift a lot when it expands and contracts as it gets wet and dry. If you live in a wet/dry climate with a heavy clay soil, you can plant gardens right next to the house. Because you'll be watering them in the dry part of the year, the soil right around the house won't dry out and shrink so much, therefore the soil will move less under your home.

If you haven't already planted deciduous vines and trees around your home, do it as soon as possible. Vegetation adds a wonderful feeling to any home. A trained vine around a window or door is a beautiful way to shade an opening from the summer sun.

Start dreaming up your gardens and next adventures.

BACKWORD

Your cob home will attract curious people from far and wide. While you are building, it's great to have a flow of potential helpers. Once you're living in your home, you may become tired of all the tourists gawking. One way to solve this problem is to have a specific day and time when visitors are welcome. If you have the energy, share your home to inspire others to build with cob.

If you are reading this after building your home, I would like to take this opportunity to congratulate you! You have created one of the most personal human rights (shelter) for yourself. I would **love to hear about your building adventures and see pictures** of the results. If you would like to contribute ideas or innovations, please do. Maybe they'll be included in another edition of this handbook.

Write to: **GROUNDWORKS**
PO Box 121
Azalea
OR 97410
U.S.A.
www.beckybee.net

All the best! Enjoy your home!!
With love, Becky Bee

BOOKS TO READ

A Pattern Language, Christopher Alexander et al. Oxford University Press, 1977.

African Canvas: The Art of West African Women, Margaret Courtney-Clarke Rizzoli, New York.

Build it with Bales A Step-by-Step Guide to Straw-bale Construction, S.O. MacDonald and Matts Myhrman, 1994, 1037 E. Linden St, Tucson AZ 85719, USA.

Building with Stone, Charles McRaven, Storey Communications, Inc. Pownal, VT.

By Nature's Design, Pat Murphy & William Neill, San Francisco Chronicle Books, 1993.

Earthen Floors, Athena and Bill Steen, The Canelo Project, HC1 Box 324, Elgin, AZ 85611.

Mooseprints, Robert Laporte, Natural House Building Center, Santa Fe, N.M.

Mud and Man: A History of Earth Building in Australia, Ted Howard, 1992, Earthbuild Publications, Melbourne, Australia.

Mud Brick and Earth Building the Chinese Way, Ron Edwards and Lin Wei-Hao, 1984, The Rams Skull Press,12 Fairyland Rd, Kuranda, Qld 4872, Australia.

Mud Space and Spirit, Gray/McCrae/ McCall, Capra Press, Santa Barbara, CA 93101. 1976.

Non-Toxic Natural and Earthwise, Debra Lynn Dadd 1993.

Painted Prayers, Women's Art in Village India, Stephen P. Huyler: Rizzoli International Publications Inc. N.Y.

Permaculture, Bill Mollison, Tagari Publications: PO Box 1, Tyalgum, Australia, 2484.

Places of the Soul: Architecture and Environmental Design as Healing Art, Chris Day Aquarian Press/ Harper Collins, 1160 Battery, San Francisco, CA 94111.

Shelter, Shelter Publications, 1973, send US $20 + $3 shipping to Home Book Service, PO Box 650, Bolinas, CA 94924.

Spectacular Vernacular: The Adobe Tradition, Jean-Louis Boupgeois and Carollee Pelos, Aperture Foundation, 20 East 23rd St., N.Y., N.Y. 10010.

The Book of Masonry Stoves/Rediscovering an Old Way of Warming, David Lyle, Brick House Publishing Co. Inc., Andover, MA 1984.

The Cobber's Companion, Michael Smith, P.O. Box 123, Cottage Grove, OR 97424.

The Earth Building Encyclopedia, (fireplace blue-print) Joseph M. Tibbits, 1989, Southwest Solaradobe School, Bosque, N.M.

The Efficient House Sourcebook, Robert Sardinsky and The Rocky Mountain Institute RMI, 1992, 1739 Snowmass Crk. Rd, Snowmass, CO 81654.

The Harris Directory: Recycled Content Building Materials, B.J Harris Stafford-Harris Inc. 508 Jose St.#913, Santa Fe, NM 87501.

The Humanure Handbook; A Guide to Composting Human Manure, J.C. Jenkins, Chelsea Green Publishing, 205 Gates Briggs Building,White River Junction, VT. USA.

The Owner Built Home, Ken Kern, 1972, Charles Scribner's Sons, New York.

Periodicals

Building with Nature, PO Box 4417, Santa Rosa, CA 95402.

Environmental Building News, RR1 PO Box 161, Brattleboro, VT 05301.

The Last Straw, PO Box 42000, Tucson, AZ 85733 (Straw bale construction).